We Are the
Middle of Forever

We Are the
Middle of Forever

INDIGENOUS VOICES
FROM TURTLE ISLAND
ON THE CHANGING EARTH

Edited by Dahr Jamail
and Stan Rushworth

THE
NEW
PRESS

NEW YORK
LONDON

Requests for permission to reproduce selections from this book should be
made through our website: https://thenewpress.com/contact.

Published in the United States by The New Press, New York, 2022
Distributed by Two Rivers Distribution

ISBN 978-1-62097-669-2 (hc)
ISBN 978-1-62097-719-4 (ebook)

CIP data is available

The New Press publishes books that promote and enrich public discussion and
understanding of the issues vital to our democracy and to a more equitable
world. These books are made possible by the enthusiasm of our readers; the
support of a committed group of donors, large and small; the collaboration
of our many partners in the independent media and the not-for-profit sector;
booksellers, who often hand-sell New Press books; librarians;
and above all by our authors.

www.thenewpress.com

Composition by Westchester Publishing Services
This book was set in Fairfield

Lannan

This book was made possible, in part, by a generous contribution
from Lannan Foundation.

The cover drawing is by Linda Yamane, Rumsien Ohlone basketweaver
and scholar, and is her rendition of one of the traditional baskets she has woven.

Printed in the United States of America

2   4   6   8   10   9   7   5   3   1

For All Our Relations

We are the middle of forever.
—John Trudell (Santee)

Contents

Preface

We often hear that "We have never been here before." A great number of species are going into extinction, we are experiencing a global pandemic, global changes in weather patterns are causing social and environmental upheavals worldwide, the Earth is on a trajectory of warming certain to cause further upheavals in both human and nonhuman communities, and scientists debate the origins of what they call the Anthropocene period, said to be the cause of all this havoc. For a great many people, the statement that "We have never been here before" means that the results of human behavior impacting Earth and the future were not apparent to them until now, yet this is not the case for all people or cultures. For the Indigenous people of the world, radical alteration of the planet, and of life itself, is a story many generations long.

For the Plains people of North America, elimination of the buffalo, from 60 million down to a few hundred, took only one human generation in the early to mid-nineteenth century. Near the turn of the twentieth century, a Plains Indian man famously said, "I know more dead than I know living." For California Indians, a greater than 90 percent population reduction occurred in the last half of the nineteenth century as a result of sanctioned settler mayhem. These are only two examples in one short time frame, examples of a vast and radical change begun centuries earlier than this one for tremendous numbers of people, wherever colonizing societies reached them. For Indigenous throughout the Americas, disease, starvation, re-

source extraction, and enslavement have taken huge tolls, and the reper-
cussions, if not the specific perpetuation of those same forces, continue to
cause radical alteration of "life as we know it." The same experience holds
all over the world.

In the search for solutions to the vast array of problems facing the global
community today, many people are looking into other cultural models for
new ideas, ideas that can also be seen as very old ways of being, ways that
were not so destructive in the long term. People search for "Indigenous Wisdom" as they imagine it to be, because with a backward glance, and most
especially through a deeper study, it is apparent that Indigenous societies
have largely had a much more integrated relationship to Earth and its
member-beings.

This collection of interviews hopes to be an aid to those looking for ideas
and responses to the conditions of today, by presenting a wide variety of
perspectives on what has happened, what continues to happen, and what
will likely occur in the near- and long-term future, given the trajectories
that surround us. Indigenous peoples have had to adapt, to persevere, to
be courageous and resourceful in the face of destruction, and this is not a
simple task. It is a very complex and long-standing effort, and much can
be learned by listening to the individual and collective human experiences
involved. In this light, the people in these pages represent many different
Indigenous cultures and communities, generations, and places within Indigenous and non-Indigenous society alike. They are telling their own stories, conveying their ideas and feelings about where we all now stand, and
each one is seeking to make a contribution to a deeper understanding of
how we may respond in the best way possible. For many Indigenous cultures, the power of story is that each listener may find themselves within
the story, and, when confronted with a crucial decision to make in their
own lives, gain from it. The story becomes an active tool, and this collection is offered in that spirit.

Stan Rushworth

Early Warnings

This story was related in 1992 by an anthropologist and archaeologist, in a class on the History of the Non-Western Peoples of North America, and it has stayed prominent in my mind since then:

At the turn of the twentieth century, as the industrial revolution swung into high gear, industrialists realized that Indian reservation lands once thought to be of little value contained the minerals, oil, and gas needed by an industrial economy. They went to the leaders of those communities, many of whom still maintained their older ways of governing, with women central in decision making, as well as listening to all concerned tribal members. The industrialists promised riches, money to move into the modern world, but the Native communities, especially the grandmothers, said, "No, this is not the right way to relate to the Mother, and there will be consequences." The industrialists then went to Congress and asked it to form a solution to this problem, which resulted in these words spoken on the floor: "We must destroy the sense of community, foster a sense of the individual, and create a good Protestant work ethic." The professor went on to describe the resulting reorganization of Native communities by the U.S. government in its own image, the formation of tribal councils with men in charge, many of whom were from boarding schools that sought to strip them of traditional values, language, and culture. Majority rule by vote often supplanted forms of consensus, and many Native lands were entered with devastating consequences that reverberate today. The grandmothers' warnings of 120 years ago are our inheritance.

The professor's narration almost thirty years ago was a rarity in academia, and was entirely invisible to the public at large. This does not mean, however, that Indigenous people have been silent about how we've come to the situation we are in today. Many have been speaking directly into their own communities and to the world community as well.

In 1944, explaining the precolonial life of her people in *Speaking of Indians*, Ella Cara Deloria wrote, "the ultimate aim of Dakota life, stripped of accessories, was quite simple: One must obey kinship rules; one must be a good relative. No Dakota who has participated in that life will dispute that. In the last analysis every other consideration was secondary—property, personal ambition, glory, good times, life itself. Without that aim and the constant struggle to attain it, the people would no longer be Dakotas in truth. They would no longer even be human. To be a good Dakota, then, was to be humanized, civilized."

Looking back on these values seventy-five years later, under a deep gray-orange sky lit by the wildfires burning all over the Western United States, in an unchecked global pandemic, in the midst of profound social unrest, including protests laden with violence and killings by both police and different protesting factions lining up against each other, with massive numbers of species going extinct in the face of expanding populations and encroachment on their habitats, and with much of the scientific world describing a rapidly escalating climate crisis that exacerbates all the social and environmental imbalances, we must consider Deloria's description of civilization honestly. By her terms, this civilization has failed. It has failed its own people, its habitat, and those whose lives it has encroached upon, Indigenous and non-Indigenous, human and nonhuman. We are engulfed in a time and state of not being good relatives, and therefore are not "humanized, civilized."

While Deloria is speaking specifically about her own people, it is fair to say that all original inhabitants of Turtle Island, now called the United States, and its neighbors, shared similar, if not exactly the same values.

This can also be seen in the early warnings of today's catastrophes by many Indigenous people worldwide. The following are only a few among many, and are those within the relatively recent past.

In 1976 Hopi elder Thomas Banyacya spoke before the United Nations Habitat Forum in Vancouver, British Columbia, in no uncertain terms. "The time has come to join in meaningful action. Destruction of all land and life is taking place and accelerating at a rapid pace. Our Native land is continuing to be torn apart and raped of its sacredness by the corporate powers of this nation." His warning is unequivocal, and his admonishment to change is equally direct. "We do have an alternative to this. Mankind has a chance to change the direction of this movement, do a roundabout turn, and move in the direction of peace, harmony, and respect for land and life. The time is right now. Later will be too late."

In 1948, a gathering of Hopi elders convened to look at the state of affairs surrounding them, and Thomas came out of that council with a message he put forward for the next half century. He was very aware that his voice was one among many, and he devoted his life to spreading this essential message to all people. His vision included all of life and was not politically centered on one people, but on all people. At that United Nations Habitat Forum in 1976, Thomas spoke clearly and strongly. "Hopi and other Native spiritual leaders are greatly concerned with the conditions of our Mother Earth. They have watched the White brothers systematically destroy the Native peoples as they did natural resources. According to our beliefs and prophecies, if this destruction continues, man's existence on this world will soon be ended. We are not asking the United Nations for help in a material way. We are, according to Hopi prophecy, simply trying to inform the world of what is going to happen if the destruction of Earth and its original peoples continues as is known by our religious Hopi elders."

He also pointed out that living in the right way takes effort, that these efforts have been going on for a very long time, and that they center on the most fundamental values. "The Hopi and all Native brothers have continually struggled in their existence to maintain harmony with the Earth and with the Universe. To the Hopi, land is sacred, and if the land is abused, the sacredness of Hopi life will disappear, and all other life as well. Land is the foundation of Hopi and all life, and the foundation of the Hopi stand."

Within his vision, there are instructions from "the Great Spirit, Massau'u" toward living in a sacred manner. "Massau'u said not to take from the Earth for destructive purposes and not to destroy living things indiscriminately." He saw the coming of the atomic bomb as the "gourd of ashes," of Hopi prophecy, which says "that many men will die, and that [when this happens] the end of the materialistic way of life is near at hand." Massau'u's instructions further say "that man was to live in harmony and maintain a good clean land for all children to come, and to take care of land and life for the Great Spirit."

Thomas spelled out that the Hopi knew this from their own experience, because we are now living in the "fourth world." "At the meeting in 1948, Hopi leaders 80, 90, even 100 years old explained that the creator made the first world in perfect balance where humans spoke one language, but humans turned away from moral and spiritual principles. They misused their spiritual powers for selfish purposes. They did not follow Nature's rules. Eventually the world was destroyed by sinking of land and separation of land by what you would call major earthquakes. Many died and only a small handful survived. Then this handful of peaceful people came into the second world. They repeated their mistakes and the world was destroyed by freezing, which you call the great Ice Age." He goes on to tell of the small numbers of people who then entered the third world, a world of technology, machines, and conveniences. "They even had spiritual powers that they used for good." But the same thing happened again. "They gradually turned away from natural laws and pursued only material things . . . while they ridiculed spiritual principles. No one stopped them from this course and the world was destroyed by a great flood that many nations still recall in their ancient history or in their religions." He said we are now in the fourth world, and that "Our world is in terrible shape again, even though the Great Spirit gave us different languages and sent us to the four corners of the world and told us to take care of the Earth and all that is in it."

However a person may approach the Hopi narrative bringing us to this fourth world, the pattern is unmistakable, of violating how we are supposed to be here, as is the necessity toward action. In this speech, he is crystal clear about this nation today, about how the instructions are being ignored. "The United States system has gone against these religious instructions and now almost destroyed all our land and our way of life. All Hopi people and other First Peoples are standing on this religious principle. Not only have Hopi struggled to care for and maintain the Earth and their existence, but the First Peoples of the Americas have struggled to maintain themselves in the world today. The present oppressive governments in the Americas continue to make the existence of the First Peoples a continuous struggle for life."

"The United States and United Nations should understand that it cannot bring about peace and harmony or the good in life if it does not correct the wrongdoings going on within the American continent."

Thomas spread this message everywhere, in large venues like that from which these passages are taken, to videos, to town hall meetings, and to people's homes throughout Turtle Island. He sought for many years to present the Hopi vision and warning to the United Nations in Geneva, and after trying three times only to be ignored, the fourth time he was finally granted ten minutes to speak, in 1992, at a time of recess, standing before a great many empty white seats. Oren Lyons of the Six Nations gave three shouts, announcing that a message of spiritual importance was to be given, then Thomas sprinkled cornmeal and spoke. The message was given, and when Thomas finished, to the applause of those few who were there, the chairman at the podium said, "I thank Mr. Thomas Banyacya. I am confident that your words and wisdom have been duly noted," then he banged his gavel once, and people filed out of the great hall.

In the 1960s, Phillip Deere (Muskogee) also traveled everywhere he could giving another clear message that "the dominating societies" were headed in the wrong direction, and that what they were doing would "unravel" before long. Like Thomas, he too went to the United Nations in Geneva, in 1977, delivering his consistent message about right relationship and what will happen through its violation. In a speech given in 1978, called "An Understood Law," he said, "I see, in the future, perhaps this civilization is coming near to the end. For that reason, we have continued with the instructions of our ancestors." He spent his life encouraging young Indian people to learn from their elders and to remember that "We are part of Nature," and that the Indian life was modeled after Nature, regardless of how hard the colonial societies tried to "make him someone else" modeled in the Western image. He explained that during the 1960s and 1970s young Indians were protesting in the streets, standing up for their rights because the ravaging of the land and water and sky was breaking their hearts. More deeply than fighting for their political rights within the colonial system, but not extraneously, they were fighting for Earth itself. They were fighting for their cultural vision, which was and is the same as fighting for Earth.

Phillip spoke of the vast differences between Indian societies and the newcomers, asking people to put social justice and environmental relationships together as one thing. "The jail houses, the prisons in this country, are no more than four hundred years old. Prior to the coming of Columbus, more than four hundred tribes, speaking different languages, having different ways, having different religions, lived here. None of these tribes had jail houses. They had no prison walls. They had no insane asylums. No country today can exist without them. Why did we not have any prisons? Because we lived by an understood law. We understood what life is all about. To this day, we are not confused."

He defined that lack of confusion as what people now term "inclusivity," but he put it differently. In describing how so many different societies managed life here together, he said, "We do not disagree on our religion. I have never tried to convert the Lakota people into Muskogee ways. We respect one another's religion. We respect one another's visions. That is our only way of existing in this country here—that is our survival. This our strength." Then he firmly added what must not be forgotten, "Even though we are greatly outnumbered, our ideas will overcome those numbers!"

Phillip talked about people following natural laws of respect, showing that the treatment of the natural world and the people of that world was one and the same, and that a society or civilization without that respect could not work, either for itself, its constituents, or its very habitat. "A confused society cannot exist forever. The first people who came here were lost. They are still lost! They have been separated from the natural way of life so far that the government doesn't understand the Indian language. People in this society have been driven away, and have been taken away so far from reality that they will not sit down under a tree and talk to us."

He also consistently talked about how the colonial people of America were "discriminated" against more than anyone, because they were robbed of knowing essential truth, both historically and philosophically. "I think the White man is most discriminated against because he is discriminated against by his own kind. The truth! We are the believers in the truth, and not the facts as this society follows." He maintained that they were robbed of a relationship to their actual history and to their own nature, both of which their civilization brought them to forget.

In 1978, hundreds of American Indians walked from San Francisco to Washington, DC, to protest violations of tribal land and water rights. It was a time when much of America was pushing back against activist movements that were asking American society to stand to its word in treaties and human relationships, and to begin respecting the moral base that Indigenous peoples throughout the Americas had and still have in relationship to Earth. Referring to the Indigenous movement, and to the Longest Walk of 1978, Phillip Deere sounded a very clear warning: "We are going to continue to walk, and walk and walk until we find freedom for all the Native people! And I will remind you, you may not be an Indian, but you better join us. Your life is at stake. Your survival depends on this."

In 1990, the Kogi people of highland Colombia reached out to the BBC to produce a documentary film to share internationally, *From the Heart of the World: The Elder Brothers' Warning*. The Kogi are a people who pulled away from the Spanish four hundred years ago, retreating into the high mountains, and they have been resisting colonization every way they can for all these generations. Although they live in the mountains, they walk down to the coast for shells used in their daily ceremonies, which gives them a comprehensive view of the results of Western civilization. This vision made them break self-imposed isolation long enough to sound their warning, then they retreated again, at that time. The Santa Marta mountains they live in are the highest mountains in the world that reach from the seashore to the tip of the mountain, from tropical lowlands to arctic mountaintops that are always below zero degrees. Their alarm is based on their observation of the death of the mountain, the drying up of the water and ice that feeds everything below it. Because they have lived in a microcosm of the planet since long before colonization, they know what is happening before their eyes.

The Kogi spent fourteen months training a man who is of both Kogi and other Indigenous descent in their ways, solely for the purposes of presenting their knowledge from every standpoint. "The Great Mother created the world in water. She makes the future in it. This is how she speaks to us. We look after Nature. We are the mamas [seers/leaders of the community] and do this here. And we mamas see that you are killing it by what you do. We can no longer repair the world. You must. You are uprooting the Earth. And we are divining to discover how to teach you to stop."

In this film, they constantly remind all people to really think, to think about all of it, to use the mind in its largest sense. They repeatedly admonish to "Think, think, think of the Great Mother," which they call the Mind inside of Nature. "At night, before you sleep . . . think what you're going to do the next day. What things need to be done and how you're going to do them. Think it through."

They explain why they made the film. "We must explain these things. We shouldn't threaten or insult, but it's good that we speak. We must show them [the Younger Brothers] how we work and how we offer our tribute to the Great Mother, so that they know that we are here working for the Younger Brother too . . . not only for the Kogis but for all the people in the world." To show the world how they care for all of life on Earth, they took the filmmaker and the translator they trained deep into their ceremonial ways, their social practices, their farming practices, and their philosophy. They opened themselves up to a world that has demonstrated callous disregard for all they hold precious and real, a courageous act motivated by their care for all of life, and by a sense of responsibility they are trying to engender in others.

One spokesperson says with clarity and authority, "I want the whole world to listen to the warning that we speak to you. She taught, the Great Mother taught, she taught us right and wrong, and still we have not given up living in the way she taught us. We remember a teaching and live by it. The Great Mother taught and taught. The Great Mother gave us what we needed to live and her teaching has not been forgotten right up to this day. We all still live by it. But now they are taking out the Mother's heart, they are digging up the ground and cutting out her liver and her insides. The Mother is being cut to pieces and stripped of everything. From their first landing they have been doing this. The Great Mother too has a mouth, eyes, and ears. They are cutting out her eyes and ears. If we lost an eye we would be sad. So the Mother too is sad, and she'll end, and the world ends if you do not stop."

Again, the Kogi beseech the outside civilization to think deeply about it all, to reflect honestly. "Younger Brother, we know the water down below has started drying up. Don't think that we are responsible. It's you. We are doing our work properly, and neither you nor I know when the world will end; isn't it so? Stop digging in the Earth. If you go on, the world will end. You are bringing the world to an end. Now we're sending this message and we ask, 'Can they think, listening to us, that it is us, the Elder Brothers, who have destroyed everything?' I know they won't think that."

To show the pragmatism of their philosophy, they teach the filmmaker how and why they farm and harvest in a way that nourishes Earth, without violating it, how they respect the balance between men and women, how important it is to love children and the world around them, how they live in the most beneficial and respectful way of touching the Earth. After showing what functions for their society, they return to the warning. "We want to give some advice, to tell the real truth to the Younger Brother, that if they go on like this and if they don't change their ways at once, that they will see what will happen, they will see what will happen."

Today, the Kogi still try to educate the surrounding world through alliances with others, through Indigenous organizations, and they fight in the courts now to protect sacred sites. They still live in cultural integrity, and they resist the despoiling every way they can. They have an ancient, pragmatic vision to share. They see the world around them, they listen to all of life, and they speak what they feel must be said.

In 1982, Mohawk leader Jake Swamp-Tekaronianeken founded the Tree of Peace Society, an outgrowth of his decades of telling the world the story of Skennenrahawi, the Peacemaker of a thousand years before. In that telling, the Peacemaker is born into a time of violence which he works to heal, as he grows older, by forming the Haudenosaunee Confederacy. In this healing, five nations who were once one, but who had broken apart, come back together. Skennenrahawi wanted a symbol of that coming together that would last, so he chose the white pine tree, which is the symbol of the Haudenosaunee people today. In the spirit of remembering, and of peace, throughout his life Jake Swamp planted trees all over Turtle Island and in many other countries of the world, directly influencing the plantings of millions of trees worldwide.

Like the Peacemakers long before him, Jake Swamp's message was a healing, and he often spoke of the grief carried by almost all people of the Nations here and worldwide, a grief born in all the destruction that has ensued, the removal of people from each other and from the natural world. His consistent message, through stories and admonitions, was to come back into right relationship, and to remember the original instructions. In the story he told, he said the people had forgotten and could not come to peace and agreement because they had no clarity to see the future. He said that people had lost their clarity because of the blinding tears of grief at the loss, from all that had happened. That pain caused tears that blinded them from seeing. And he said that people had lost the ability to listen because of the grief felt at the pain from what had happened. He said people had lost their voices from the grief and pain, so they could not come to agreement. The abilities to see, to listen, and to speak are the foundations of how to communicate and thereby find a future together.

He said the people needed to take the stricken ones by the hand, and help them to "raise their eyes to Creation" and find "the purest cloud to wipe away the tears," for vision to be restored. Then they should look to Creation again and find "the softest feather for opening the ears," so they could "hear the wind, the birds, and all the things that make sound in the world." Then they should reach to Creation still again and "find the purest water to wash away the lump in the throat," so they could speak again. To see, to listen, and to speak clearly are the essentials that brought agreement to his people a thousand years ago, and now we need this again. This is his message.

To accomplish what he believed was needed, he made himself available to all, to large or small groups, through planting ceremonies or talking circles or discussions or conferences, whatever was needed. He was known everywhere as a kind man who included everyone. Mohawk artist Billy Myers said of Jake, "He did not close the door on anyone, Indian or non-Indian." Mohawk elder Leonard Four Hawks said, "He was able to transcend any culture, any nationality, by planting trees. More importantly, he followed the oldest tradition that we have. Whatever we do today will affect the seven generations to come, which is 140 years from now. We don't live long as humans, nor do most animals, but trees do." When Billy Myers planted six white pines in honor of Jake one year after his walking on, he hung "spirit ribbons—black, white, yellow and red to represent the human races." He was following Jake's lead, as Jake Swamp included everyone.

In so doing, he represented the foundations of Indigenous thought we hope these pages bring readers into feeling and knowing. He was one voice among many, one who reached many people worldwide, yet he did not restrict himself to a broad audience alone, but was consistent in his message no matter the situation. His children's book, *Giving Thanks: A Native American Good Morning Message*, was a gift to each individual child who learned its essential prayer, giving each and every one the strength to create a good life, in gratitude. Its opening prayer is: "To be a human being is an honor, and we offer thanksgiving for all the gifts of life." In keeping with all the speakers in this collection, past and present, this is what must be remembered.

Stan Rushworth

The Interviews: Editors' Note

The conversations of this collection are in the order of their occurrence. We did not organize by theme, as many of the themes brought up by the people overlap, which they have to when considering the all-encompassing change they are addressing, as well as the peoples' tenure on this land. Although we have fundamental questions that will become apparent to the reader, we felt it was most important to follow each person's lead in bringing their thoughts, feelings, and experience to the forefront. This manner of approach has allowed us to learn from each autonomous individual, and from the many traditions out of which they grow and live. Each has given us tools of understanding, of thought, of greater connection, and we offer these perspectives in the aspiration that they may provide the same for you.

In keeping with clear presentation of very different voices, experiences, thoughts, and feelings, each person interviewed has had the opportunity to read and edit their own part of the collection. We trust that this process provides a representation of each individual involved with the greatest amount of integrity possible. We appreciate the kindness of each person that contributed to this offering, and it's our hope that the views gathered here may be helpful to all who read them. Thank you for reading.

Dahr Jamail
Stan Rushworth

We Are the
Middle of Forever

1

President Fawn Sharp (*Quinault*)*
Strength

COMPOSED BY DAHR JAMAIL

Ultimately the solution to the crisis lies in our values, and we've proven that simply by existing today, regardless of how we had the most powerful country in the world try to destroy us, terminate us, and assimilate us. We lived under great pain and suffering. They carried out murder and genocide and attempted full-scale annihilation, but they never could stop that drumbeat in our heart. One could either just wither away like paper, or be like steel that just grows stronger and stronger. When the most powerful country in four hundred years can't stop you, you know it is because of our resources, prayers, and blessings, and everything that has been across this land since time began. And we not only have survived, but we are now emerging even stronger.

—President Fawn Sharp

"These are the homes of our elders," Pierre said, standing on the beach of Taholah, Washington, pointing at the gray, weathered wood siding of a small home standing barely fifty feet from the surf crashing nearby, with

---

* Since this interview took place, President Sharp has been elected to vice president of the Quinault Nation during the Quinault Indian Nation General Council Meeting on March 27, 2021. At the time of this writing, President Sharp is president of the National Congress of American Indians.

a couple of others just like it not far away. Out of respect, the Quinault give their elders their best home sites upon which to live.

The only thing separating these rickety structures from the coastline, which is ever advancing from sea level rise and increasingly potent storms, is a small barrier of sand dunes and boulders that were placed there as a breakwater. But even that has long since become half buried by sand.

Pierre Augare is the special assistant to the president of the Quinault Indian Nation (QIN), Fawn Sharp. He was showing me around the reservation while I awaited my time with the president.

"Obviously we'll be moving these houses uphill, which is where we've already cleared a site for relocating this village," he continued, then pointed up above to nearby hills covered in western red cedar, Douglas fir, western hemlock, lodgepole pine, and Sitka spruce, some of them reaching nearly three hundred feet high.

The Quinault are among a very small number of Native Americans who live and hunt on the same land, and paddle across the same waters as their ancestors did centuries and centuries ago. The Quinault and Queets tribes comprise the Quinault Indian Nation, along with the descendants of five other coastal tribes, which include the Quileute, Hoh, Chehalis, Chinook, and Cowlitz.

The ancestors of the modern Quinault lived in a way that shared in the cultures of the people living both north and south of them. Subsisting on sea mammals and massive salmon runs, hunting wildlife, and harvesting from abundant forests provided more than enough of the physical and spiritual necessities for their ancestors. The western red cedar is their "tree of life," as it provided logs for their oceangoing canoes, bark used in clothing, and boards for their longhouses.

The Quinault are the Canoe People, and the people of the cedar tree. "We remember our past while employing modern principles in a marriage that will bring hope and promise to our people now and in the future," reads the tribe's website.

President Sharp, having been told by her elders to run for president, obeyed, won, and had been in office since 2006. And not needing to be told of the folly it would be to attempt to brace or push back against the most vast, deepest, wildest ocean on Earth, within days she began enacting her plans to move the QIN's two villages to a site a hundred feet above sea level.

Pierre continued our tour, showing me the tribe's small gym, community center, a small grocery store, and multiple other buildings and homes that were all awaiting relocation to higher ground.

While he did, I couldn't help but think of my hometown, Port Townsend, located on the other side of Washington's Olympic Peninsula. With its small downtown at roughly six feet of elevation and right above the water, it is in the same situation as the villages of the Quinault, as well as dozens of other major coastal cities around the planet that will either be relocated entirely, or swallowed whole by the sea. Many of the inhabitants of Port Townsend take pride in being politically progressive, with the vast majority of residents being acutely aware of the climate crisis that is upon us.

Nevertheless, the best the city could come up with as a response was pouring millions of taxpayer dollars into tearing up downtown's main street, aptly named Water Street, to upgrade electrical and sewage infrastructure underneath it in a project that was completed in 2018.

Pierre took me to the uphill site for the village. The QIN's hospital is already there, as well as the main road that is going to run through the area. Trees have been cleared to make way for what is to come. Pierre told me how, literally, everything we had just seen down near the crystal blue surf is to be moved uphill.

Born in 1970, Fawn Sharp graduated from college when she was nineteen years old. Five years later she became an alumna of the University of Washington School of Law, after which she returned to the QIN to practice law for over a decade as their tribal court judge.

During that time the previous tribal president was preparing to retire. Sharp's elders asked her to run for his position, but she was reluctant. Having trained to seek truth, justice, and fairness, she saw the role as political and she had no interest whatsoever in becoming involved with politics. "The idea of being a politician did not reconcile with my personal concept of how I wanted to devote my life," she told me in her office at the tribe's headquarters, where we spoke after my tour of the reservation.

We sat in her office across a wooden table from one another as the summer sun shone outside. Photos of loved ones lined her windowsill, and rows of law books filled a bookshelf along a wall. The Quinault tribal emblem, a wooden carving in the shape of a large eagle feather, and a carved canoe paddle, hung on the wall behind her.

Out of what she described as "a sense of duty in response to the elders," President Sharp continued to take steps towards the presidency, and announced she would seek the office. But she was hesitant, and weary of what she saw as a potential morass of politics. At just that time, an elder pulled her aside and she explained her aversion to politics to him. "He said, 'Look, you're not running to be a politician, you're going to be a leader, and a leader brings those virtues to office,'" President Sharp said. "'That is the difference.'" Hence she learned her first lesson in public service before she was even elected: "It is critically important to ensure throughout your service that you hold the office and the responsibilities in high regard."

She sees her role as tribal president as a "very sacred role." Bringing Indigenous traditions with her position of power, once elected she immediately began the process of decentralizing power from the presidential office and enacting a community-driven agenda. After being elected in March 2006, President Sharp also went to work immediately on restoration of their declining runs of blueback sockeye salmon.

The blueback are seen as the most prized of all the sockeye, as their rich oil content makes them a far tastier fish. The Dutch word for the blueback translates as "excellence." The Quinault used to harvest from runs of millions of blueback in the 1950s, but since then they have been in steep decline. The ensuing decades saw the runs drop from the millions to the hundreds of thousands, to fifty thousand, to, in the last few years, President Sharp said, "It's just been a blip." In 2018, she explained, the QIN harvested twenty-seven sockeye. "So we closed our blueback fishery," she said, after I asked her to clarify the number as being only twenty-seven fish.

That was her second introduction to the dramatic impacts of climate disruption, knowing that the climate disruption–fueled warming waters of the Pacific Ocean are the major contributing factor toward declining salmon runs. The tribe's scientists showed her overlays of ocean temperature graphs with the salmon decline, and the temperatures nearly perfectly mirrored the declining salmon runs. They talked with her about ocean acidification. Then they talked with her about melting glaciers, which led to her first direct physical experience with the impacts of climate disruption. It was this that made her, literally, physically sick.

President Sharp was taken on a helicopter flight over the Anderson Glacier in nearby Olympic National Park. "We were hoping to see the glacier, and it was in that moment we discovered it was completely gone," she said somberly. She leaned toward me from across the table and continued softly, but sternly: "It had disappeared. That was quite a moment for me. It was then when it hit me at a deep level."

In October 2018 she took another flight to see if the glacier had come back at all. It had not. Additionally, President Sharp noted how much the nearby Eel Glacier had receded from her previous flight. Retreating 10 meters each year, and at a rate that is accelerating, the Eel had shrunk roughly 1,000 feet.

"So I am coming face to face with all of this, after a decade of struggling and fighting," President Sharp said, her voice strained, nearly cracking with emotion. "It really took an emotional toll on me. Some of our team saw me get out of the helicopter. They said it looked like I was about ready to throw up."

Yet things were about to intensify further for her, as though she were being forged by the same fire that is threatening Earth.

Not long after that experience, President Sharp took another heavy hit. Ballot measure 1631, an attempt to pass a carbon tax in Washington State which she had worked tirelessly to pass, failed to do so as the result of the fossil fuel industry spending $33 million to defeat it.

President Sharp had so internalized the defeat that she left the country for Mexico for three days, in an attempt to catch her breath. But, upon returning to the United States, she peered down from the airliner onto the record-breaking wildfires in California during the season that the town of Paradise was incinerated.

"I'd always heard of the psychological impacts of the climate crisis," she told me somberly. "But I don't think I'd really felt it until this last year, after all of those events."

Yet it was the culmination of the blueback restoration project, her efforts to move their villages to higher ground, and numerous other projects related to the climate crisis, underscored by that series of shocking events, that led President Sharp to an understanding of the gravity of the role the Indigenous have in responding to the climate crisis. She had come a long way in her work, given that during her first term, she addressed the crisis while receiving much criticism from within the tribe from people who felt there were other more pressing issues to be dealt with.

"So it took a while," she continued. "And it's taken a lot of deliberate effort to connect the climate crisis with things on the ground, but over the last decade our people have come to realize what is happening. And now here we are and Washington State has just declared another drought."

She was a young president with no political experience, finding out immediately that the biggest priority within her community was the loss of an iconic salmon species that not only represented the food they have relied upon for centuries, but is part of their identity and an aspect of everything they do, from births to weddings to funerals. Despite having held leadership positions such as being a trustee of the Washington State Bar Association–Indian Law Section and vice president and founding member of the National Intertribal Tax Alliance, and having received a degree in International Human Rights Law at Oxford University, President Sharp had (and continues to have) her work cut out for her.

When she had brought the climate issue up during intertribal, state, and federal meetings during the last two years of the Bush administration, she was faced with the challenge of engaging even a single person in the climate conversation. Hence, she reacted by, in 2008, attending the COP 14 climate summit in Poland, having already had two and a half years to pull together her agenda around the climate crisis. "The goal was to pull fifty-seven tribes in our five-state region, along with Canadian First Nations people, into a land base that would collectively be larger than the European Union," she said. Since the United States was not a signatory nation of the Kyoto Protocol, yet an Indian tribe could be, her goal was to pull all of the tribes in her region (and Canada) together in order to open up a discussion in Poland about the carbon cap-and-trade market. But, before she arrived, countries had already begun backing off from advancing the cap-and-trade idea, in what would come as yet another defeat Sharp would have to overcome.

This trend continued into the Obama administration, when President Sharp watched then-president Barack Obama's envoy attend the climate summit in Bonn, Germany, then conclude that the United States lacked the political will to take a stronger stand on the crisis. Undeterred, President Sharp, along with other tribal leaders, hosted a UN meeting at the National Museum of the American Indian, which found other countries that were very open to dialogue and advocating the interests of Indigenous peoples. She believed the agenda the Quinault had created was worth taking to an international stage, despite the lack of response from the U.S. government, and despite having had to declare a state of emergency early on during her first term, when in December 2007 a storm of nearly hurricane strength besieged the Quinault. It was a wake-up call, as their power was knocked out for eight days, and there was no water in what she described as "an epic event." She has had to declare four national states of emergency thus far, and is doing so without a completed climate mitigation plan.

President Sharp had hoped that by working with Washington State governor Gregoire, and then with Governor Jay Inslee (with whom she was engaged as part of his carbon emission reduction task force), she could assist in advancing climate policy for the state. However, with these efforts having little effect, "It made me realize if we can't achieve climate policy in a state like Washington with leaders like Inslee and Gregoire, we are in serious trouble."

Thus, four years ago she began pulling together an idea of taking climate policy directly to citizens, due to the fact that she believes the average person understands that the climate crisis is real. "They [the general public] understand the fossil fuel industry needs to be held accountable, so I thought in the Trump backlash there could be a sort of political wave, and in the 2018 election, and within Washington State we would be able to become the first state to put a price on carbon," she added. But she underestimated the extent to which the industry was willing to go in spending more money to defeat measure 1631 than had ever been spent on a Washington State ballot measure, as the total price tag for the campaigns from both sides was more than $50 million.

"I actually made a statement at our last general council meeting that we may be the last generation to know what bluebacks are," she continued, her tone becoming increasingly resolute. "I thought long and hard after I got back from Mexico, and decided that I wanted the next chapter of our climate agenda to be even more aggressive."

Now, the QIN's official policy objective is to make the fossil fuel industry wish that ballot measure 1631 had passed, because, according to President Sharp, "That was a bargain basement price of $15 per metric ton of carbon for them to have to pay. So our six-point climate agenda that I introduced a few weeks ago is something that we'll now be addressing even more aggressively. I have a duty, and know there is a crisis, and it's not only a crisis facing the Quinault, but people all across the world. We are left with no choice but to now come at the fossil fuel industry even harder."

The QIN is a sovereign nation with the inherent right to govern itself and deal with other tribes and nations alike on a government-to-government basis. The tribe's website states: "After 150 years of mismanagement by the federal government, it was obvious that tribes could manage their own affairs better and make their own decisions without external interference. This is the basic underlying philosophy of Self-Governance."

Hence, in 1988 the QIN's Self-Governance Act began as a "demonstration project" in the Bureau of Indian Affairs (BIA). Then, in 1990, the QIN, along with six other tribes, implemented self-rule in Indian affairs. Today, their tribal operations include Natural Resources, Health and Social Services, the Quinault Beach Resort, and Community Services, among others, all of which are geared toward promoting the growth and development of the full potential of their reservation.

"It may take another century to correct the many problems created by the 'Indian agents' we once relied upon," states the QIN website. "But we now look to the future while learning from the past."

President Sharp is extremely clear about the fact that, despite all the challenges the tribe has faced over the decades and centuries, "We never relinquished our spiritual connection to the land and ocean, which is way more powerful than a piece of paper that speaks to ownership," she said, enunciating the words slowly and clearly.

Their 208,000-acre reservation, located on the remote southwestern corner of the Olympic Peninsula, contains lush Pacific Northwest forest, strong-flowing rivers, emerald-blue lakes, and nearly two dozen miles of raw, largely untouched Pacific coastline. The QIN people share their home with cougar, bald eagle, black bear, elk, blacktail deer, among other animals, and the nearly 4,000-acre Lake Quinault is not far from the coast.

President Sharp told me of her fourteen-year-old son who has been gifted eleven songs from the lake that, as she said, "have not been heard for over a hundred years. He told me that Lake Quinault is reawakening in order to heal herself."

Her son had been pressuring her to learn more about their culture and traditions. "He told me that if I'm the president, I need to know all of this history and our legends," she said, laughing. "He said, 'You should know who you are.' My response to him was to tell him that my generation had to go and get an education and become accomplished in this world, so that

I could come home and support his generation. But that it is his generation that is now getting to learn our language while they are children."

President Sharp said that her son's generation is the seventh generation
since first contact in her area, for which she is deeply grateful, because
that means the Quinault did not allow seven generations to pass without
language and traditions being passed forward; otherwise their language
would have been lost.

A cultural revitalization movement in the Pacific Northwest, which began in the 1970s, has continued to gain momentum with time. A large part
of this has been the annual canoe journey that takes place each summer.
Tribes from around the Salish Sea send delegations in dugout canoes from
their tribal communities to a designated location, usually a host nation of the
Coastal Salish people, where they all converge. The event has grown to
include more than one hundred traditional oceangoing canoes from Indigenous nations from as far south as Oregon, and as far north as Alaska.
The journey can last from two to three weeks, and is now a huge event each
year for President Sharp and the QIN.

"We didn't allow seven generations to pass without being back on the
canoes," she added. "So his pressuring me has helped me a lot. Last year I
was in the canoe journey, and was asked to stand and ask permission to
come ashore. I felt like a baby, learning all of this; it felt like I was crawl-
ing and it was my first tribal journey, and I'm just now learning our ways."

Realizing that, she told her son, "You guys are the leaders. I'm here to
support you."

President Sharp understands the power and importance of the blessings
and ceremonies that have existed on the land of her people for centuries.

"Anything that happened in the last four hundred years really pales in
comparison to the power of the connectivity of the songs across the country,"
she explained. "I explain to people that no matter where you go within tribal
communities you are going to hear a drum. It is the heartbeat of Native
America. And in my mind it does not have a beginning and it does not have
an end. It's just one of those things that is created. No matter what public
policies were leveled against us over the years, nothing could break our

spirit. Nothing could stop the drumbeat. Nothing could break the connection that we feel at a cellular level, having our whole being level with the natural world and the environment."

President Sharp discussed environmental activists, and how she sees most of them engaged in what she referred to as "a very mental struggle," then said, "But for us, it's just who we are and what we do. I cannot imagine not advocating for the natural world in these circumstances. It's just part of our teachings and it's part of that continuation."

She had felt that connection when she was out on the canoe journey the year before we spoke. It was when she truly experienced it firsthand. She hadn't physically trained her body for the arduous journey in their dugout canoe, but when it launched from nearby Honshu Point into fourteen-foot swells and she could barely see their support boat, the time to live the old ways was upon her, ready or not.

"I remember thinking 'There's no way I'm going to be able to do this,'" she told me softly. "Twenty minutes into the paddling, we were barely past the rocks, and I was already physically exhausted. But our skipper started singing one of the old songs, and once he sang that song, it was as though the ocean came to life and we were dancing with the ocean through the song. I felt this incredible, almost superhuman, power."

She had already told her young people that if they look at the landscape, "That's the same landscape our ancestors saw for centuries. There's no development, no hotels, it's just pure and closed to the public." It's this that she remembered while canoeing, as the strength was sung back into her body through the ocean that was being sung through the song.

"That experience reinforced at a spiritual level what I knew at an intellectual level," she continued. "During my first year as president, I was invited, during a television interview, to come up with a myth and a truth. One that came to mind through prayer was the myth that the Europeans believed upon first contact that we were primitive and we were savages. But then there is the truth that if you look at scholars and scientists and people like Abraham Maslow and the hierarchy of maturity, at the very base are selfish people, then as you climb up you get to the independent people, then finally the interdependent people."

President Sharp went on to point out how someone who is self-confident becomes independent, but only when they go on to care for other people have they arrived at the point at which they are considered to be a highly mature individual. This is the truth she went on to share in contrast to the myth.

"I relate this to how we as a people [Native Americans] were not only interdependent relative to our fellow humanity, but we were interdependent relative to the natural world . . . to the animals, to the trees, to our Creator, to the Great Spirit that lives in everything," she explained. "That is what Chief Seattle referred to, that all things are connected. What we do to the Earth, we do to ourselves. We are just one strand in this intricately woven fabric."

This is what came into her during the canoe journey, among the massive swells as the song was sung and the ocean danced and the strength flowed in and through all of it. She both experienced it, and thus "knew" it.

"Then to witness my son rediscovering these songs that have been lost! The first one came to him when he picked up an eagle feather, and he said, 'Mom, when I picked up that feather, it's like the song just came to me.' Another one was while he was at recess from school. He goes from Seabrook here to South Beach, and that was one of our trade routes where we had trails, and they would stop there next to one of the creeks. And it was at that spot another song came to him. So it's very powerful and it's very real."

This is what made her realize the importance of Native nations and tribes during the climate crisis. This is a time when tribes are beginning to occupy a leadership void caused by the absence of federal leadership on the climate crisis. Now the average citizen is beginning to understand and recognize the power of the treaties. "They seem to be the last line of defense against fossil fuel exploration in this country," she said. "The Lummi defending Cherry Point against coal, or those who are fighting against the crude oil exports, or Standing Rock."

It all connects for her, as President Sharp is watching Quinault children understand what Indian country is about. She remembers the television commercial of the Native American seeing a garbage dump, and the camera pans in to show a tear running down his cheek. Native children are realizing the value of our shared waters, contrasted with the idea of isolationism or superiority that is so prevalent in the dominant culture.

"We fight against this because we are embattled, and we struggle, because we are in opposition to all of this, and we have so much adversity, but we really just want to live in peace with people, and be at peace with our world," she said softly, as tears welled in her eyes. "So we advocate for better climate policy because we are aware of the global crisis, and how this will require everyone, because no one is immune from the climate crisis, and everybody has a responsibility."

She paused, then went on to state that everyone has their place, that we just need to remember the North Star of our values, and specifically for Native Americans, the precontact lessons that are timeless and have been proven through the centuries and millennia, and more recently, as science has proven what the Indigenous have always known—that they already had the best practices with Earth.

"NASA did a whole study on science and traditional ecological knowledge, and their conclusion was that what we have always done was also the best science," President Sharp explained. "And I know if our ancestors were alive today they would still be doing what they do, which always have been the best practices."

She believes tribes can lead the way in transitioning from fossil fuel use to a better, saner way of living. While most people understand how Indigenous communities are already the first and worst impacted by the climate crisis despite having had the least to do with it, what many non-Indigenous people may not know is how deeply connected Indigenous people were, and are, to the areas where they lived, and still live.

"For our ancestors, this area was part of them, and now it is part of us," President Sharp said. "So when the Earth suffers, we suffer. In contrast, if someone moves to another area, there's no foundational rock or any sort of connection to it that they value and appreciate. So when we see the climate crisis, there is a sense of responsibility, and an inherent quality of feeling a deep sense of responsibility to seven generations out."

In October 2019 Sharp was elected as the twenty-third president of the National Congress of American Indians, perhaps in part because of how she lives her sense of responsibility to the future generations wherever she goes.

"When I leave the reservation or fly to Washington, DC, and I'm looking at the landscape, I no longer see it as sort of foreign soil. I see it as part of a rich history with an unbroken chain of prayers and blessings from when time began, and I feel the ceremonies," she explained. "And that will always echo throughout eternity. Nothing can kill or destroy that spirit, and that is who we are."

Acknowledging how the average person in the United States feels powerless and questions if the country is even a functional democracy, she reminded us how the real power in the country does not lie in the Oval Office, or the White House, or Wall Street, but in the voice of the engaged citizen, Indigenous and non-Indigenous alike.

"Historically, when we face crises, *that* is when people come together and rise above that conflict and embrace true values, *then* it becomes our finest hour," President Sharp said, speaking in a softer, lower voice, leaning forward on the table. "Right now our generation is faced with a challenge, and with or without anyone else, we are going to strive for this moment in time to be our finest hour."

President Sharp paused, took a deep breath, then continued.

"I think our Creator has a perfect plan that is in perfect timing for everything. We are led through times of crisis for a greater purpose that we'll never understand until later. So our goal and objective is just to be true to Creator's planning and calling for our life, and to resist the temptation to become apathetic or negative, and just be true to our own purpose. Because we are all here by design, and with a good heart and good intentions we can constantly seek each day the wisdom and guidance we need."

President Sharp believes that in every sector of society leaders exist that will rise above all the negativity and look to the greater good.

"We must resist falling into an ideology, or politics, because we just don't have time for that, and the work at hand is so serious; there is so much riding and dependent upon the work we do. We have to face each day with prayer and seeking guidance because this crisis truly does exceed the scope of human understanding. But so do the solutions."

President Sharp leaned back, took a long breath, looked out the window, then at the photos of her family members and children in the room, then back at me. Speaking calmly, yet resolutely, she concluded:

"Ultimately the solution to the crisis lies in our values, and we've proven that simply by existing today, regardless of how we had the most powerful country in the world try to destroy us, terminate us, and assimilate us. We lived under great pain and suffering. They carried out murder and genocide and attempted full-scale annihilation, but they never could stop that drumbeat in our heart. One could either just wither away like paper, or be like steel that just grows stronger and stronger. When the most powerful country in four hundred years can't stop you, you know it is because of our resources, prayers, and blessings and everything that has been across this land since time began. And we not only have survived, but we are now emerging even stronger."

2

Gregg Castro (*Salinan/Ohlone*)
A Sense of Permanence

COMPOSED BY STAN RUSHWORTH

What are you supposed to do? You turn to your stories, you turn to your oral narratives, you turn to your learning, you turn to your Traditional Ecological Knowledge, you turn to what your culture has given you and taught you to do, and that tells you what you're going to do.

—Gregg Castro

By his own declaration in *News from Native California*, Gregg Castro is a zombie, one of the walking dead. He is a former tribal chair of the Salinan people, a t'rowt'raahl Salinan/rumsien and ramaytush Ohlone, and an active participant in the Society for California Archaeology for twenty-five years, advocating for respectful treatment of Indigenous sacred sites. Gregg also speaks at local colleges and universities, and is interviewed by radio and television programs shedding light on local Native history and ongoing issues. He is a storyteller at Native gatherings, and a tireless advocate for as many Native peoples' rights as possible. Despite an articulate, consistent, and kind presence, he is invisible in the sense that his people have no reservation, no federal or state or local recognition, no existence as defined by the American culture surrounding them. At the same time, Gregg can drive an hour south of where he lives in the San Francisco Bay Area to visit the place of origin of his people, a mountain that now carries a name not its own, and only his people still say the original name aloud. He is a Native man who can sit on a spot on a mountainside and look back in time for at least fifteen thousand years and see a long chain of ancestors working on that very spot, chatting, dancing, praying, and living in a way they had evolved for a very long time, a way that worked well for them and their surroundings. Gregg can sit and see his relatives, and he can feel their presence, and when he speaks, this presence is always with him. He calls it "a sense of permanence."

Talking about his childhood, Gregg says, "Even though my dad was raised in a family that raised him to know who he was, it wasn't safe to tell others." He recalls that during the Mexican period, the Gold Rush period, and for a generation after, it was safer to identify as Mexican than Native, as this could save one's life. Referring to his contemporaries, to many Native youth, and to those of his parents' generation, he says, "They grew up not knowing who they were . . ." because for "a lot of our people, their family and tribal history was suppressed." This makes Gregg's work, and the work of a great many people like him, essential today. They carry a huge responsibility because they carry an antidote to generations of that suppression. "We weren't ashamed, we were proud. We knew where our homeland was, and we lived close to it. Our family still lives in this Salinas/Monterey area," and "there were a lot of Indians there. Having that growing up, I think that gave me an anchor that others didn't have." There were too many who had been forced to forget.

After World War II, families spread out to find work, and his went north to San Jose, "but I still had that connection because we'd go back often." Part of his homeland is "in the Los Padres National Forest, and part is on the Hunter Liggett military reservation," but to Gregg, "it's still looking like the place the ancestors would remember from thousands of years ago. If you ignore the bombed out old trucks used for target practice, it still looks like it did a thousand years ago. Walking those trails, I see the rock mortars that my ancestors used for thousands of years," and "I've seen them my entire life," and this forms "a literal rock-solid connection." This connection is what he passes on to the young and to others searching for a sense of connection to place, land, and culture.

This is not an easy task. As a kid, he understood his root connection deeply, but when he went to school, it changed. "Objective knowledge came from the outside. They told me, 'Well, the California Indians are dead, they're all extinct.'" When he told his dad this, his dad said, "Well, if you're dead, they won't try to kill you." Gregg calls this a "very pragmatic" stance, and chuckles. "In my family, we didn't have shame, but we might have had fear, the fear that it could go bad again." The culture surrounding him talked about the land and world and his people differently than he did. "It wasn't until later when I grew older that I began to understand that other people didn't have that. They'd talk about their long history in the area, and I'd say, 'Oh yeah, really?' And they'd say, 'Yeah, yeah, we've been here a hundred years.' 'Oh, okay.' I just knew we'd been here since the beginning of time."

Gregg would ask himself, "How does an Indian person who's supposed to be dead, extinct, fit in with this? It was the two opposing ideas, and feeling that pressure to reconcile them; that was the thing that was most unnerving to me, because I always knew who I was, but I had all these people telling me I couldn't possibly be that because I'm dead. I'm extinct. I don't exist." Some Lakota in-laws came to California and showed how deep the erasure of his people was. "Even they didn't realize at first that we were still around. And so there was this oddity, of two opposing worlds, and that's the sort of life I've lived, not in confusion, but conflict, maybe, of two worlds butting against each other, and me in the middle." But Gregg's base is deep and strong regardless, because his father "was always very clear about who he was, and that's who we were. What's to talk about? It was never an issue for him."

The same solidity he describes in his father carries through in him without any compromise.

I ask Gregg to speak about climate disruption and the surrounding society's response. "Well, some of them think, 'Oh, no, everything's fine.' They think their God is going to fix it for them, so they can do whatever they want. That's a typical two-year-old, in a sandbox being a bully. And in their short understanding of their existence, their version of permanence is eternity to them . . ." but "it's not eternity. Our stories tell us, very specifically tell us, there was a time before us. There was a time of Creation. There was a time when the first people, the real first people, were here taking care of the place, shaping it, forming it, taking care of it, to prepare it for us," and "a lot of the stories talk about us being last. Not first, we were the last. And it was an incredibly beautiful place."

He describes how the last child in a family has everything in place already. "Our stories tell us what was given to us, and what we're supposed to do with this." He talks of Dr. Darryl Babe Wilson, an Iss/Aw'te culture bearer, who said, "When we come last into the world for us to be in, all we have to do is learn how to take care of it, and be grateful for it, and humility and gratitude are some of the most basic thoughts that we should be having, because we didn't do anything to earn this." Gregg says, "This may be one of the most fundamental differences between cultures because they came in thinking 'We came in God's image,'" while "we are taught we need to know our place," and "the society that's grown up around us is still throwing tantrums in the sandbox, still thinking the universe owes them everything. It's extremely immature," and "it's a huge immature baby that's fully capable of destroying itself.

"And that's the other part of it that our stories tell us, that there isn't always a happy ending. There's no fairy tale. Our stories tell us there are consequences. That's what they're there for. They talk about various people, creatures, beings who screw up," and "sometimes they don't survive their screwups. And that's the way of life, and two-year-olds don't want to hear that. The two-year-olds can't understand, let alone see, their own mortality and the possibility of it."

Regarding the environment and climate, Gregg says, "The elders have been telling us for a very long time, 'There's a big problem here,' with major issues in what we're seeing in the landscape, in the weather, in the animals, in the trees, in the atmosphere, and it's really bad, and they've been ignored. And it's hard to make it right now because people are still looking for the two-hour Hallmark movie version to make it right, that we might have some hardships, but in the end the sun will break through and we'll all hold hands and we'll be fine. And, there's a great possibility that this won't happen.

"We have end of the world stories. And it doesn't end well. In some cases it gets completely remade. The world survives, but everything in it does not. It gets transformed, and life begins anew, maybe with a whole new set of players who'll get it right this time, and maybe they'll survive." Gregg pauses, then adds, "I don't think people understand the danger we're in, and just how fragile the world is." He restates his thought, with emphasis now on the word "world": "The world itself is not fragile, it's fragile for people."

I ask Gregg how and why he thinks we have gotten to this point.

"Even the most forward-thinking of them are still human-centered. 'How do we save the world for us?' they ask. When we sing, the Earth enjoys that, and it gives us a blessing because of that, but it doesn't need it. We need it. And now we have it upside down. We're at the top, at the pinnacle. But no, we're at the bottom. If humans want to survive, it's not too late, but it's going to be on a drastically different level." He says we have to shift.

He talks about a history class, where history begins with Columbus, and the thousands of years before that don't really exist in the mind. "That is a constrained way of looking at things for their historical purposes. Well, they tend to look at environments that way too, in a constrained view." But, "We are a blink in the world's eye. Indeed, we have it upside down."

I ask how we might put it right side up, and Gregg offers that, "Optimally, we have to disabuse ourselves of the idea of wealth. What does being wealthy really mean? We were incredibly wealthy before the Europeans showed up, because we had everything we needed. There was no homelessness. Nobody was starving. Everything was there for us, and we considered that to be wealth. A lot of wealth. Enough wealth." In Native gatherings in his childhood, "Everyone was welcome, even though from different groups, including Whites, because they're people. We treated them as equals, and I think kids are seeing that today." He points out that young people trying to adjust to "radical wealth disparity" is one source of a shift in thinking. He talks briefly about "tiny houses," and other shifts in how young people perceive what is really needed. "I think the question is, will it happen fast enough?"

He jokes about TEK, Traditional Ecological Knowledge. "Now that they can give us an acronym, they might begin to talk to us," and "there is some science, politics, and economics we could share with them of 'the before time' that would be helpful, but before any of that can happen, it's got to happen right here in their hearts, and in their minds. Their frame of reference has to change, and I don't know if there's a trick to that.

"It's a mind-set, a heart-set. Darryl Wilson said it's a dawning realization of how totally alien our ancestors thought. It wasn't human-centric. It wasn't self-centric. It wasn't about us. Our languages don't even have that many words for 'I.' It's 'we.' A lot of our words are pluralities. That's the thing that's got to happen first; 'we' . . . that basic understanding that it's a 'we.' I'm not sure I have an idea how that could happen, how to change that. . . ."

Speaking further on TEK, and how it might be approached, including what stands in the way of beneficial change, Gregg points out that "They're still worshipping their religion, and their religion is science." With decades working in Silicon Valley, he is familiar with the viewpoint. "I understand some of it, the mechanics, the math, and all of that data. And that's what they're talking about when they talk about TEK. They're looking for the secret formula that's going to save them. And it's not a formula. It's a change of your entire life and outlook on the universe. It's much more fundamental and transformative, and if they keep looking for a set of formulas, they're never going to get it. That to me is the fundamental problem. They're still trying to see the world with the very limited science that they have before them, that they worship." He smiles and shakes his head.

"Darryl talks about the difference between the brain and the mind. Mind is far beyond the brain. It's a view of the universe that's not constrained by science, and as intricate as modern science is, it's also very limited." Nonetheless, he also emphatically says that the methods of science "are just tools, and you can use tools in different ways. If it's not working, you need to figure out how to reimplement that tool in a different way that will accomplish the goal." He does not close the door on it, but qualifies science's purpose and position. He says his science is Indigenous, based on what works for the total system of life. It's the pragmatism of his father, and of his people, and his enduring sense of permanence despite the enormity of change going on.

"One of the issues that many Native people see with Western science is that science is only good if it is detached, if it's objective. 'They' have this idea that in order to understand something you have to break it down into tiny components," but again, "like a two-year-old, they forget how to put it back together." They say, "You have to eliminate human emotion from the equation. You have to be detached, objective, an observer, not involved, not connected, but now through various theories things are connected. That's the next step for them." This can imply an ethical shift, but "their breakthrough is still Eurocentric. It is not new, lots of cultures have it," but do they recognize this? "And is it going to be in time? . . . Yeah, we're connected. It's a brilliant idea, and it's an idea that's long overdue, and it's a question, again, of whether it'll happen fast enough."

I ask Gregg the question I've heard numerous elders deal with again and again, for we must remember that Indigenous people have been dealing with destruction of the world as they know it for a very long time. If we question that we might not have time to turn things around, how do we then carry ourselves? And how does that form what we do?

"Darryl talked about a part of the story when you cross over to the spirit world. You keep on going, and the Milky Way is the path you can follow, and our ancestors are there, but you can keep on going. They didn't think of 'planets,' but they saw them as distant fires, and you could go there, and they understood that those were places to be, places where, well, there were people. You could go there and learn. You could go across the sky and guess what, you could come back, if you want, and you don't necessarily have to come back as people, and there was nothing wrong with that, again, because our philosophy isn't that we're better than anybody else. It's that everybody's equally valued here, in terms of life value. All creatures have the same right to be here as us. None of this nonsense of humans as the superior being. We're proving quite well right now that we're not the superior being, because we're destroying everything. That's not the mark of a superior being. So you can come back as something else, and there's nothing wrong with that. So, that also tells me that if we're not here, and some other being dominates, so what? First of all, it's not our position to say: it's the Creator's. And if we contributed to that, then it's our fault. Not our decision, but our fault. And there's nothing to say that can't happen.

"In the meantime though, we're just talking about our corporeal bodies. But our spirit, in a lot of the philosophies of Native people, is part of Creator, so it's eternal, and infinite. We are just a tiny speck of the infinite and eternal. But it keeps on going. We can have fun on those other planets. Hopefully we won't mess those planets up, and we'll keep going, and maybe by the time we get back, we'll have learned something, for the second go-around. And we won't screw it up so bad. Or at all! But this idea that we're the be-all and end-all . . .

"We're a speck on the timeline in the world's history, and as a society we just can't grasp that perspective, and we refuse to, which is why we're still doing stupid things. If we really understood this, we'd stop doing the silly things we're doing now. And, we could stop this." Here is where the pragmatism of the old story can take hold, if it is listened to.

"Yes, we could fix it, but are we going to?" So far, Gregg says, "I don't see any movement of the mind, the frame of mind, the philosophies, the understanding of people, because the rest is just mechanics once we have that understanding. . . ."

There is a period of silence, then he repeats, "It's still salvageable. But it will take a massive . . ." he trails off, waits for words, then speaks strong and clear. "It'll take Darryl's dream. Remember when Darryl talked about it? He told the young people to go out in the morning and sing the sun up, and this advice had a deep effect on them." Gregg smiles. "The song in the morning by the children, on the mountain top. That's what it's going to take, some spiritual equivalent of that, if not that exact thing, is what's going to have to happen."

Within the philosophical shift he talks about, there are also the mechanics of cultural preservation and renewal, of retaining this vision. For twenty-five years, Gregg has worked to protect sacred sites and "remains," and the odds have not been in his favor. He says the system is currently set to protect the views of those newer to the land, but he can't let that stop him. So these are the functional mechanics, the kinds of attitudes he shares, the persistence that comes from being rooted in the land forever.

"We talk about how 90 percent of our ancestors died, and 90 percent of our cultural sites are gone, and in all the time I've been working in archaeology, they've still been eradicated by modern society and its destructive ways. They keep doing it. They create these laws that are supposed to protect sites, but they put so many holes in them, like Swiss cheese. That's the purpose. That's what the laws are about." With this he adds, "So why am I still here?!"

He points out that over 90 percent of the time, Native peoples' efforts are in vain, that it's a perfect mathematical parallel, a continuity of the same process, but he still works as hard as he can, and this points to the methods of Indigenous resistance to what many others are now aware they themselves are facing.

"I once asked Darryl, 'Why are you doing this [telling the old stories in classrooms, writing, speaking, recording stories, and teaching]?' And he said, 'Because it's the right thing to do.' This is an alien twist that our contemporary society doesn't understand, where there is a hierarchy. There has to be a winner and a loser, and a large part of our Native societies aren't concerned with any of that. It's about, 'What are you supposed to be doing right now?' You have these tools, you have these capabilities, you have this opportunity. What are you supposed to do? You turn to your stories, you turn to your oral narratives, you turn to your learning, you turn to your TEK, you turn to what your culture has given you and taught you to do, and that tells you what you're going to do. Maybe not the fine details, but if you've been paying attention, it gives you enough to do the right thing."

Here Gregg gets to the core of it for me, the heart of what the surrounding world needs to hear. "It's my privilege, and my honor, my sacred obligation and duty, as a person gifted by my culture, enriched by this culture, and this is how I pay it back. It has nothing to do with the batting average, which is horrible. It has to do with that every single time I have an opportunity, I need to stand up, and it doesn't matter the consequences. That's what's so radical about our philosophies. We're not so concerned with consequences."

He talks about the corruption that happens within systems that have consequence only at their forefront, then he returns to Native philosophy. "When you look at a deeper way of looking at the world and your proper place in it, which is what our ancestors tell us to do, you're coming from a place of humility and gratefulness, because it's a gift, every moment of our life and every molecule of our life is a gift. We didn't earn a damned thing, so why are we being such jerks about it?"

Contemplating his words and his annoyance, and the steady patience that quickly follows it, I have to come back to the focus that never leaves me. I ask him, "How do you see that 'sacred obligation' Darryl and others talk about, especially to the children?"

"That's a challenge. It's more than a challenge. We don't grow up in a community like in the old days, like in a village, where we saw each other every day, and how we did or did not live out our ideals, our philosophies, our sacred obligations, and that doesn't happen any more in the way our social structure is. Our kids are out there influenced by many forces. But, our elders tell us we can still learn the older TEK. 'Get out there and sit under a tree for a while. Look at your homeland. Look at the ground. Take off your shoes. Stick your hand in the ground . . . listen, because it's talking to you.'

"Darryl told me a story: He remembered being swaddled up in a cradle board. 'You're a little cocoon. You ain't going nowhere.' He remembered being bound in, either stuck in the ground 'like a cornstalk, or hooked over a branch.' He remembered hanging from a tree. 'You're just there, and the tree branch moves in the wind, and you can't do anything. You're there for the ride. The board is turning a little bit, so you see a different part of the world as it turns. All you got is your eyes, and your mouth to breathe, and that's all, and it gives you a different outlook,' and I thought about it, and how these elders have a different way about them, a different way of interacting with the world. Maybe they didn't have cradle boards, but they had the spiritual equivalent of it.

"Darryl talked about how that, rather than being a bad thing, was a really awesome thing. It teaches you your place in the world, not in a bad way and not to make you feel small, but that you're a part of it. And you are there, absorbing all this beauty in the world, and you don't do this in isolation. You see your family working, as the board turns in the breeze. Even in the stories of how the world came to be and how we came into it, you learn that you're in 'the web of life.' You're a part of it.

"When we say you're a small part of the world, people can think, 'Well, I'm diminished, of less worth,' but that's not what 'small' means. That's not what it should mean. It just means the universe is huge, and you're a part of it; you're an intimate, integral part of it.

"Even the small children could go out and pick up acorns, and that's important. Maybe you could only pick up a couple, if you're a small baby learning to walk, or if you're an elder who can barely walk, but you're a significant part of the community. That's the vital part, and that's what the ceremonies are about, like when you cross over [pass on]. It's acknowledging the importance of you to the community."

To some, this might be a picture of a community that is impossible to form in today's world, but to Gregg, it is a functional social system that builds responsibility on as many fronts as possible. "I don't think our ancestors were all that saintly, but I think they had a beautiful, wonderful philosophy, one that many of our kids don't understand, and that's our job. That's our sacred obligation, to help them to understand that it's a whole other way to look at the world.

"Our ancestors, after tens of thousands of years, had a beautiful philosophy of life that extended beyond our limited sight and understanding, and they had confidence in that. That's real faith to me. The most incredibly deeply faithful spiritual people I've ever been around are Indian people, not religious leaders and priests, but no, elders, because they know it with every fiber of their being, and then they live it. Not perfectly, but well enough. That's what our sacred obligation is, one that we're not doing a good job of, for all kinds of reasons, some well beyond us. We must pass on that beautiful alien philosophy that our ancestors gifted us with, that way of living we haven't been able to translate.

"But it has to happen for everybody, perpetrator and victim. And their descendants, because it's a nasty virus we're living with." He tells the story of an insect bite on his wrist, one he kept ignoring until it formed a quickly growing boil full of bacteria, spreading like a virus. "I let it go; I didn't think it was a big deal, but when I went to the doctor, he took one look and said, 'There's only one thing to do,' and he grabbed a scalpel. 'You have to lance this and get the poison out, because if you don't, it's going to go through the entire system.'

"And what's this got to do with climate change? Again, you're not going to fix anything else until you fix yourself, because one thing that'll happen is that we will find ourselves here again. Let's say someone does come up with that miracle box in Palo Alto, and sucks up all the pollution, and we go back to where we were a hundred years ago, and they take their billions and billions of dollars they're going to make, and create another crisis for us to fix, because we haven't learned a damned thing? That's the problem.

"That's what we're dealing with now. We're not going to fix anything, including climate, until we fix ourselves and fix this diseased society we're in. I live in San Jose, and I know what it's like. I drive up El Camino Real every day to go to work, and I've been doing this for three-and-a-half decades, and in the last eight years, there's been a massive growth of tents all along the road, right along the railroad tracks. That's what modern greed has done. In the relatively recent history of 'civilization,' greed is the disease, as Darryl used to talk about. Greed and 'disrespects' have poisoned the entire body."

A long silence follows Gregg's words.

I finally ask him if he has anything to add. It's late, very dark outside, with owls calling back and forth in the pine trees, and he thinks awhile before speaking again.

"I try to stay motivated, in various strange ways, because if you look up and look at the horizon, how far off it is, and then you see the horizon's on fire, you lose hope. I try to make sure, whenever I go to the homeland, to stop at the graveyard at the San Antonio Mission, because that's where four thousand of my ancestors are buried. They died building that mission, and I stand there and I look at them and I sing to them, and I talk to them, so that I can remember, that whatever I go through pales in comparison to what they suffered in just one day, with massive death on a scale we can't even imagine; and the suffering, and I imagine the loss of hope."

A big silence rises between us, then he speaks again. "I guess I'm not an optimistic kind of guy. I'm a realistic guy, but I like to think I have hope. And it's based on fact. And the fact is that up to 95 percent of our ancestors in this state died, during a little over a hundred years, after being here since the dawn of time. The dawn of time! In a hundred years we almost went down, and yet, here we are! It's hope based on reality. If they could survive that, we can survive this."

There is more silence, more owl song, then Gregg says, "Survival may not be what we think it is, but you know what? I'm not afraid of that anymore. I'm not afraid of that because that's one of the gifts that I've learned over the years.

"And that's what I would like to share with kids: Don't be afraid. Don't be afraid. You can be nervous, but don't be afraid. That fear, that deep-seated fear, it just gets in the way. And if I can, I share with my kids and young people that 'This too shall pass,' and most important, that we don't know what's going to happen, but that in the moment, in this moment, there is something you can do. You do have control over you! That's what the message is from our Creator and our ancestors, that 'You can deal with *you*,' and that's all the control you really have, and that's what your obligation is to think about, right here, right now.

"And that's a big enough package as it is. So, focus on that. If you got that, things will fall into place. But you have to understand it's a package deal. There's a lot that comes with that, because you are talking about controlling yourself, but in this you're also affecting others, and you're affecting others by what you do, and if you really care about those people, you're going to want to do something good, so that you affect them in a good way. That should be the basics of life."

Gregg pauses, then finishes, making sure I know the sources of his thoughts and feelings. "I'm just passing this along. I've been blessed, incredibly blessed, by elders and knowledge bearers who took the time to share with me, and I only have to say I took the time to listen, and that a lot of people still have that opportunity. I happened to be at the right place at the right time, and I listened.

"And now when I travel across the state, I see an elder, and I know. I know 'That's a culture bearer,' and they're not wearing any labels, but I know. And it's not for sale, because it's a gift, and they do it in little ways." And they are still here on all the different homelands, still here.

"So you watch, you listen, and you learn. That's the gift they give, and you have to listen and look. And understand."

That is the key. And it is the admonition of the many "walking dead" among us, those who are proof we can rise to any task, if only we have the will.

3

Ilarion Merculieff (*Unangan*)
Living from the Heart

COMPOSED BY DAHR JAMAIL

And this is the way my people lived; fully embodied trust without thought. We embodied this faith in our lives, ourselves, in Mother Earth and the Universe, and in the Great Spirit that lives in all things. And this is a way that was the way of the original human beings that used to promulgate throughout the entire world, and we have forgotten that.

—Ilarion Merculieff

Born and raised on St. Paul Island in the Bering Sea, Ilarion Merculieff was raised in the traditional Unangan way. Since then, for more than five decades, he has worked with the traditional knowledge, wisdom, and spirituality he has acquired and shared with culture bearers around the planet. Ilarion has acted as a bridge from a much older culture to those of us living today by way of his cultural role as Kuuyux, a traditional messenger for the Unangan people.

Ilarion lived half his life on St. Paul Island, and has lived in Alaska for his entire life. Throughout his life, his passion has been serving Indigenous rights and wisdom, speaking and teaching to promote a harmonious relationship with Mother Earth. Ilarion's work has taken him down many paths. Currently the president of the Global Center for Leadership and Lifeways, he was the first Alaska Native commissioner of the

Alaska Department of Commerce and Economic Development, a state cabinet post.

Ilarion has given keynote addresses at the National Academy of Sciences Annual Meeting and the White House Conference on the Oceans, and chaired the Indigenous knowledge sessions of the Global Summit of Indigenous Peoples on Climate Change in 2009. The awards and positions he has held are too numerous to list here, save mentioning that Ilarion won the Environmental Excellence Award for lifetime achievement from the Alaska Forum on the Environment, and is the author of multiple books and essays, including his book *Wisdom Keeper: One Man's Journey to Honor the Untold Story of the Unangan People.*

Ilarion's current passion is speaking of traditional elder wisdom and its application to modern challenges, such as the climate crisis and COVID-19, but he has always had a particular focus on restoring women to their place as our original healers, and balancing the dysfunctional masculine dynamics that are so prevalent today.

Ilarion greeted us with a gentle smile as he sat at his desk in his home, a drum leaning against the wall behind him. He spoke very slowly and deliberately throughout our time with him. "Aang Waan," he said in Unangan, before explaining it was their way of saying "Hello, my other self." "We say that to each other every day in greeting each other."

Ilarion made it a point not to refer to himself, nor his people, as Aleut, "because that's the name given to us by our former oppressors," and Unangan means "the people by the sea."

His traditional name is Kuuyux, which was given to him when he was four years old by the previous Kuuyux. The meaning of the name is "an arm extending out from the body." Ilarion explained that he, as Kuuyux, is a carrier of ancient messages into modern times, as a messenger. "So now I'm living the legacy of my name."

The Unangan have a remarkable history as a people. They trace their roots back to Egypt, including migration to Outer Mongolia, to Kamchatka, then across the Bering Sea by skin boat to Alaska. The Unangan were always steeped in spirituality. Ilarion pointed out that they were the only

Native people of Alaska who didn't have footwear, even in the winter. "You know how they have stories about the Lamas who sit on snow and melt it," he commented. "Well, that was my people."

They did not have any food storage technologies, except for air-drying, yet the Unangan developed the most densely populated linear mile of shoreline in all of North America, this at a place where it was challenging to eke out a living. But an elder had told Ilarion to watch the birds, as they don't worry about where they are going to get their food tomorrow. "They just are," he explained. "And this is the way my people lived; fully embodied trust without thought. We embodied this faith in our lives, ourselves, in Mother Earth and the universe, and in the Great Spirit that lives in all things.

"And this is a way that was the way of the original human beings that used to promulgate throughout the entire world, and we have forgotten that," he added.

Ilarion was of the last generation of Unangan to have had a fully intact traditional upbringing, on St. Paul Island, which is twelve miles long and five miles wide. When Ilarion was a child, the 450 Unangan people there shared their small island with 1.2 million northern fur seals, 2.5 million seabirds, and a thousand reindeer. "It was a very magical place to be raised," he commented.

When he was four years old, to get to know his grandfather, and his grandfather him, he lived with him full time for two years, including going to work with his grandfather. "I went to visit his friends with him while he drank tea, and they would talk. I would pray with him where we'd go down to the Bering Sea. We would take our shirts off and spray ourselves with the salt water, praying toward the east. Then we would just walk. We've lived there for more than ten thousand years, and we're still there. Sometimes at night we might go to the Russian Orthodox Church, because he didn't see any difference in where one got one's spirituality. He was very much a traditional elder, and he taught me probably the most significant thing in my life."

That "most significant thing" was this: St. Paul Island is referred to as the birthplace of the winds. "That's where the winds start, before they go to the mainland," he said. "So it is very unusual to get sunny days. We get twenty days of sunshine a year, so it's always overcast or foggy."

But on the day Ilarion spoke of, it was clear and there wasn't a cloud in the sky. The sea was calm, he couldn't hear any waves, the seals were barking, and the birds were singing. He said to his grandfather, "It is such a beautiful day." His grandfather said one word, "Listen."

"That single word, I learned as I got older, encapsulated why our people survived and thrived in the Bering Sea for over ten thousand years. And the most important things in life must not be reduced to words because if you do that you confine it. You shrink the meaning of what you experience by giving it words."

In his traditional upbringing, his people didn't use many words to express themselves, and Ilarion said the best way he can describe the way they were was "very present in the moment and in the heart. Through that place of the heart, it tells the mind what to do, not the mind telling the heart what to do."

According to Ilarion, Yupik elders of Southwest Alaska call our so-called modern society the "reverse society," or the "inside out society," because it has reversed the laws for living. This is because one of the most salient laws is that the heart used to tell the mind what to do, and the mind's job is to figure out how to implement what the heart is telling you.

Ilarion sees the great imbalance currently playing out across the planet as stemming from the fact that "We left our hearts. And when we left our hearts, we left ourselves."

His upbringing fostered the opposite of this. "Literally the entire village raised me, and I had to spend equal time with the men, the women, the elders, and my peers." The men taught him hunting and fishing and the men's ways, and the women taught him food preparation and the women's ways, while the elders did none of those things. The elders, instead, told him story after story, from when he was five years old until he was thirteen. "I would just be glued to listening to their stories, and it was just so rich," he explained.

In that world, it was the adults' job to create the space for the child to learn, but not to tell him what to learn, how to learn, or to define anything. "So I grew up not asking a single question," he said. "I did what we were expected to do, which was to watch, listen, and learn, and that's what I did."

Furthermore, he was never scolded. Instead, he was always affirmed by every person in the village that he would encounter every day. "'Good boy.' They would always say something like that," he said. "And I could go to anybody's house, day or night, and be greeted like the long lost son. I was told, 'Come in, sit down, eat.' That's the way I was greeted by every household anytime I would go there. And it was through that way of learning that I was able to gain the meaning of my name, and to carry these messages all over the world."

And he has been doing exactly that now for more than half a century.

Ilarion shared another story, this one from when he was six years old. He was lamenting the fact that his people had forgotten how to make ceremonial masks. He sought out the oldest man in the village, who was eighty-eight years of age, and whose grandfather "was the last shaman of the village." Ilarion continued, "I told him of my lament, and he said, 'No, it's never been lost. You want to get it back? You go out to the beach, you take a stick and a rock and bang them together to the rhythm of the ocean, the rhythm of the wind, rhythm of the seals and the birds and the grass. You just get it to the point where you just find that rhythm and then when you find it, you don't have any thought in your head. And go to your center, your heart center, and set your intention and wait.'"

Ilarion explained that his elder meant, literally, to get out of his head, and that was what he did, and he's known how to do that since he was six. "I learned that from the seabirds, by the way," he added. "I got from my head into my heart."

He learned to set his intention without thinking, by literally embodying faith.

"To trust that you set your intention and wait means you must have the trust that the answer will come," Ilarion said. "So I did that and waited for two or three hours, and then all of a sudden a black dot appeared in my mind's eye and got bigger and bigger and out of it poured like a hundred different masks. I'm not an artist. I draw stick people. [He laughs.] I thought, 'Well, I must be imagining this,' right? So I go and tell the old man, Aggy [his teacher] was his name, and he said, 'You touched the womb at the center of the universe.'"

Ilarion explained that that womb is why women were considered sacred all over the world at one point in time. His teacher explained to him that the womb at the center of the universe is where we get all of our things that we live with each day, including our original instructions.

"It was not something that we invented over time," Ilarion explained. "It's something that was given to us. The original instructions were identical all over the world, except for the specific culture; that came from the language that's used, which comes from the vibration of the land, and then people create their own version of the same original instruction."

Women have identical energy fields to the womb at the center of the universe, so, Ilarion said, when they get together in sisterhood, they have the capacity to clear out years of intergenerational trauma, and can do so very quickly, whereas men take upwards of a lifetime to get to something we need to heal.

"Once women do that, they focus their energy ceremonially in creating that same energetic space, that same field of great awareness outside of themselves," he explained. "Then something new will be birthed. And until that happens, there's nothing new in the world that's going to be created. Nothing."

Ilarion used this grounding to gather Indigenous elders from around the world in Kauai in 2017 to discuss two questions: What is the state of the world as they see it now, and what must we be doing now?

He described the experience as magical, and shared how there was lightning and thunder and rain every day, all day long, except for when the elders went to the ocean and held ceremony. And as soon as the ceremony was done, they had rain, thunder, and lightning again for the entire day.

That happened every day for four days of ceremony. "That stopping of the rain, thunder, and lightning was affirmation that what we were doing was in harmony with the rest of the universe," Ilarion said. "And we had many, many such things happen like that during that time."

They created what he called "an organism, not an organization," of like-minded "heart people" who would work to implement what the elders wanted, which was to get their message out to the world, at which point a Reuters reporter approached Ilarion and asked how he could help.

On Earth Day, with help from that reporter, a short film of the gathering and their message was launched internationally. "This is not something that we did," he said. "We did the work we had to do, but the rest is not up to us. It's up to the Great Spirit that lives in all things. So we don't concern ourselves with the number of people that we target, as long as our intentions are in harmony with the universe and with the Great Spirit that lives in all things." The fourteen-minute film, with the messages of the elders, was shared in fifteen different languages.

The message that came through was what the elders were saying, very much aligned with Ilarion's own cultural teaching: "We need to go to the heart and be in the heart, because the heart tells the mind what to do, and will guide us impeccably. Because it understands and exercises love and compassion and understanding and patience and all these kinds of things that we strive for."

The elders told him that we must let go of all our human attachments and go into the heart center, which is the center of the river of life, because "it knows where it is going and we don't, and we need to find others who have done the same," he explained about what they told him. "Those who had the courage to jump like you did."

The elders told him that this is going to be the new definition of tribe, and the old definition is going to fade away.

"The elders are saying there's very little time left," Ilarion said, speaking to the gravity of the converging crises besetting the planet. "And that time is even shorter for human beings. Mother Earth, she has survived for billions of years. She's going to survive for billions more. It's a question about whether or not we, human beings, are going to survive during this time. And it's going to be decided by those in this lifetime, who are alive today in the world."

Ilarion reminded us of how Chief Arvol Looking Horse, the nineteenth-generation Keeper of the White Buffalo Calf Pipe Bundle, who is spiritual leader of the Lakota, Dakota, and Nakota people, said, "Do you think you were brought here now by accident?"

"The Great Spirit that lives in all things knows that you are here at this time for this purpose, and so the elders say you have a gift to give to the world," Ilarion said. "To the world, not to yourself, but to the world. And the only place that you're going to find this gift is in your heart."

Ilarion found his gift roughly fifty years ago, but during a time when he said he was "at the peak of patriarchy." It was a time when he held the cabinet post of the commissioner of the Department of Commerce, Community, and Economic Development for the state of Alaska. He was vice chairman of a $3 billion company, and headed up seventeen boards and councils.

"I gave it all up in one day," he said. "Because I knew that my time with this system was done, and that I had to give it up based on what my people have shown about what we need to do, which is to just simply be in the moment and in the heart and that the rest is taken care of."

Thus, he released the logic and rationality that the dominant culture tells us we have to use in order to think of the future and make plans.

"I didn't plan my career, for the first time in my life. And I didn't know where I was going to live. I didn't know what I was going to do. I didn't have any money. I didn't know anything. And I just surrendered it all, and people thought I had really lost it and had gone crazy." He laughed.

He "took the leap," he said, and very shortly thereafter, "that's when the magic in my life happened."

He was invited to visit friends in Alberta, Canada. While there, Ilarion was invited to a sacred ceremony on the Morley First Nations Settlement that hadn't been performed in 150 years. An integral part of the ceremony was when elders arranged four sacred pipes in the four directions.

"The thing was, the smoke didn't go up from the pipes. It went sideways and the smokes wove together a weave that went straight up. And I knew then that these were true elders. So I watched, listened, and learned."

In the middle of the ceremony, an elder spokesman said they knew why Ilarion was there, and not just because he was invited. They had been praying for someone to carry their message, and the message was that they knew that many people think they had forgotten their spiritual ways.

"We want you to know that that is not true," he said of what they told him. "It has been kept for you, and the unseen world is waiting for you to wake up in spirit."

The experience affected Ilarion deeply, and he earnestly began looking deeply into himself, toward the Great Spirit, and the Spirit that lives in all things, for guidance toward what he was to do.

After the ceremony, he received a call from someone, also in Canada, who said she was a messenger for the Hopi and Maori people. They met and she shared the story of four sacred stone tablets that contained their laws for living and their prophecies. He agreed to carry the messages of the tablets, even though he didn't know anything about it, nor had he spoken to large groups of people.

Shortly thereafter, upon his return to Alaska, he was invited to emcee an Indigo Girls concert, during which he decided to share what he learned in Canada to a crowd of four thousand. Since then, he's been carrying messages like this and others to the world, because the elders he works with are telling him the world is looking in the wrong direction for answers to crises like the climate crisis, and COVID-19.

"People are reacting to situations around the world, be it corruption, violation of women, wars, murders, suicides, refugees . . . any of these issues. And the elders said that what you choose to focus on becomes your reality. So if you're choosing to focus on stopping something, in reaction to what somebody else did, then what you're doing is pouring your mental, physical, and spiritual energy into that. That makes that thing grow because your energy is going into it. And they said that instead, you must look at what world you want to see, not in reaction to anything. Just the world you want to see and to do that, you have to go to your heart."

This is the core of the message the Indigenous elders he's spoken with around the world want him to bring forward: be present in the moment, go into your heart for the answers to all the questions you might have, and the answers will be provided.

Ilarion then shared his thoughts on the global pandemic.

"With COVID-19, Mother Earth is speaking. She has been speaking a long time to Indigenous people and she's crying for her Earth children. She's trying to help us wake up, and since we haven't listened, has introduced this virus. It comes from Mother Earth to teach us to slow down."

Ilarion used the example of how in Alaska, with more than two hundred villages and two hundred tribes, one thing that is very distinct in all of them is that they walk, talk, and act slowly. "They know what they are doing," he said. "We call it the Earth-based pace. They're slowing down to move at the rate of hearing Mother Earth. Going faster disconnects us from our relationship to Mother Earth. Another thing that it teaches is that we must take stock of what we have done."

The brief reprieve from the frenetic pace of the industrialized world, due to the COVID-19 temporary economic global shutdown, caused skies to clear, rivers to begin to heal, and animals to return to places populated by humans, for the first time in generations. All of this happened in an extremely short period of time.

"She's showing how quickly She can recover from the damage that was done if we are to just listen, to take stock of what we have wrought, and go in a different direction. But the world unfortunately is not listening. And so the elders are saying that the next things that are going to happen will be worse."

One elder told Ilarion that the people who will be best suited for dealing with these increasingly challenging times that are upon us are the "people of fire," meaning people of the heart.

"This man told me we must not *think* love. We must not *do* love. We must *become* love. Every single human being on Mother Earth has this charge, and some are going to listen. Some are not going to listen. And how it's going to turn out is anybody's guess right now."

Stan, having taught large numbers of college students, some of whom are interested in pursuing careers in the sciences, asked Ilarion to share his thoughts on whether it is possible for science and Indigenous wisdom to exist in a good way together. Ilarion, who had immersed himself in Western science for thirty years with the aim of bringing traditional knowledge and wisdom to Western science, concluded then as he does now, that science lacks humility.

"Because of that, they [policy makers using only Western science] talk about incorporating traditional knowledge into Western science, and they don't acknowledge the part wisdom plays," Ilarion explained. "But when I met with the elders in Alaska about this, the Western system calls it subsistence, and they have all their laws about subsistence to govern us, and they said that is the wrong way to go."

Western science will continue apace, but Ilarion said the elders believe Indigenous people must develop the capacity for different tribes and regions to come together as a people and figure out, together, what must be shared with the world. But right now, Indigenous perspectives are largely invisible to those adhering strictly to the perspective of Western science.

"Because of this invisibility, it is still a struggle for Native people all over the world to get their way of knowing into the decision making that affects their lives."

He chaired Indigenous knowledge sessions for the Global Summit on Climate Change, where Native people from eighty nations gathered to talk about the climate crisis.

"They concluded unanimously that governments will not act fast enough and that we don't have climate change as they understand it," Ilarion said, reflecting on the lack of urgency he was seeing in the Western culture regarding the crisis. But he was careful to differentiate between urgency and panic/fear.

"The elders, they don't like to foment worry or fear," he said somberly. "So they don't talk about the thing that I'm going to say, which is that the human species may have, maybe, five years left to decide whether or not we're going to survive."

Ilarion continued, speaking slowly, even more deliberately.

"I'm going to be the last Kuuyux left alive amongst my people from thousands of years of the Kuuyux tradition. I'm going to be the last one either because we are dead, we are gone, in which case there's no sense in trying to have a new Kuuyux, or there's going to be a fundamental change in consciousness in human beings, then a new Kuuyux is not going to be needed anyway. So either way, I'm going to be the last Kuuyux. There are many traditions that have messengers like me that say the same thing. Business as usual is not going to work. It's not going to work. And we're all scurrying around trying to maintain businesses as usual around the world, and these solutions are not going to work. We have got to change our consciousness, and we have got to change it now."

Ilarion changed the direction of the conversation then, lightening the mood. He spoke of people who are being born, and who are amongst us now, who already know deep spiritual truths that would take most of us a lifetime to learn.

"They already know these things by the time they are ten, fifteen, or twenty years old," he explained. "They just know it. And we don't understand it. We think it's just a fluke, but it's not a fluke, because they're here for a purpose."

To underscore this, Ilarion shared a story of meeting a five-year-old elder in Alaska at a gathering of tribal elders on the Kenai Peninsula that he emceed. The gathering had to be outdoors, as there was no money to rent a facility.

"There were hundreds of people there, and there were thick black clouds throughout the region, so we knew it was going to rain. We knew we'd have to cancel it if it was going to rain because we didn't have money for the facility. But then this five-year-old asks for a drum. Somebody gives him a drum and he sings a song while he hits the drum. And while he was singing, the clouds split in half and went in the opposite directions, and it was blue sky the whole day. This elder also speaks four languages fluently, and no one ever taught him."

Stan mentioned how two of our interviewees for the book are young women from his classes, both of whom had very difficult things happen in their lives, yet carried a depth of love for the world that was beyond words.

"Sometimes we're given these tragic things in our lives to wake us up sooner because there is no time left, and it's no surprise they would have to deal with real life challenges that most people would have trouble negotiating," Ilarion responded. He went on to tell us how he prays for all the homeless people he sees around Anchorage, and thanks them for being here, and for being human, and having the courage they have to have in order to be here at this time. "Because, you know, according to these elders there are no more gurus. We are all gurus now."

In his work and talks, Ilarion has often spoken of the duty of men to protect sacred space for women so they can do their work. I asked him to speak of how this can be done, especially in Western society. He explained that there are as many ways to do this as there are people, as there is no single way or formula, and the need would differ depending on the location as well.

Ilarion told another story. While facilitating a gathering of women at the Sea of Galilee in Israel, he explained to the men the need to protect the space of the women so they could concentrate on healing themselves and creating a container of energy, and pouring that love out to the waters.

"While we were doing that, we had a fire, and there was a man who was the fire keeper who felt like he had to tend to the fire while the women were doing the same," he explained. "And of course this was a no-no, because it was invading the space where women were doing their work to tend to his fire. It created such a stir amongst all the women that were there. The men need to protect the space, not invade the space of the women, and so that means he should not have done that."

Using himself as another example, given he's a man often speaking about women's issues, one of Ilarion's women teachers told him it would take a man to open up the door so women could do their work.

Another example is the powwow.

"When you see the powwow, the women are always in the back, dignified and dancing very slowly. The men, on the other hand, are dancing like chickens; they're all over the place and doing their thing. But the women are calm, cool, and collected. They are setting the tone and the space, the sacred space for the men to do what they do, because the women are already connected with the Great Spirit. They don't need to do all that work, yet the men have to work way harder at it."

He points out how women's work is now usually considered invisible, despite the fact that it was always the original sacred teaching.

"Today women's work is invisible and is taken advantage of by those in the imbalanced cultures. Yet this work is more important. I think that the women who understand these ways and do them, they are very important during this time. According to the elders that I work with, it's actually the most important work that needs to be done, because nothing new is going to be created on Mother Earth until that happens."

Ilarion explained, as he had before, that this is because women have the identical field of energy as the womb at the center of the universe, which is the place of creation and creativity. "They have that within themselves," he said. "That's why women were considered sacred all over the world."

Ilarion has also spoken of how languages are made by the land where the people live, given how Earth's vibrations form the cultures and languages of those who live where they do, assuming they are listening, so he returns to this idea, wanting to make sure it's thought about deeply.

"This is something linguists don't pay attention to, and that's the vibration of the language. The vibration of that language is what gives us the ability to talk with Mother Earth in the locale where we live. For example, amongst my people, we have words that cannot be explained in English. When we hunt for Stellar sea lions, we work to kill the animal in the first shot. Sometimes it doesn't happen, and the sea lion starts to go in a circle, in its own blood. And we say, 'Rundeedukt,' and that word means multiple things."

The first meaning is that the animal is preparing to die. The second is that it is performing the ceremony for it to die, and third, it is giving itself up. "It means so many different things, and that's just one example of this vibration that our people picked up from the place [in which] they are living," he said.

Ilarion pointed out how he comes from a tradition where he can travel anywhere in the world, be with Indigenous people and understand what they're doing, even when he cannot understand their language or the culture. "But I can feel the vibration. And this is very, very important, but most people don't think about this when it comes to saving Indigenous languages. They contain the very code that is necessary for them to talk with Mother Earth and for this exchange to happen."

He thought for a moment, then mentioned other ways of being able to do this, such as "going to the heart, being in the moment, and trusting." He kept coming back to this teaching for a reason. He wanted us to hear it, and to really consider its ramifications.

I asked Ilarion to expand on what he'd spoken of regarding the importance of listening. "What is it, as you see it, that has kept the bulk of Western society from listening," I asked.

"I think the answer is very simple, but it's hard to get to," he said. Ilarion spoke to how, when he goes to restaurants in cities, "I have to holler to be heard because it's so noisy in the restaurant. Noise is something that we are taking for granted in cities. And most people are starting to live in cities now. And so they're ingrained in this thing of noise."

As he sees it, one of the few opportunities people in cities have to communicate, or to be part of, and to show respect for Mother Earth, is during meals.

"When we eat, we don't talk very much; it's not a time to talk. It's a time to appreciate and have reverence for the fact that Mother Earth is providing these wonderful, nutritious life-giving forms that we consume. But that's not ever mentioned anywhere in terms of food preparation. Spirituality is a word that is not understood by most people in the world, I think. Then they don't slow down. Mother Earth is showing us we need to slow down in order to talk with Her. And so we have to slow down, but what's happening in the world is that each generation is talking faster and faster."

From Ilarion's perspective, everything is accelerating, and the ubiquitous acceleration of everything is seen by the dominant culture as a good thing.

"But again, the Yupik elders say that we are living in the reverse society because we reversed all the laws for living, and we are doing exactly the opposite of what we need to be doing."

I asked him if he thought that some of this is, at least in part, a symptom of people's fear of really listening to the heart, and of living in their hearts; hence, the noise and speed is essentially an escape from following their own hearts.

"Exactly. Everything is based on the fear of the unknown, and fear disappears when you are in the heart. The heart does not worry about what's happening tomorrow or yesterday. It just is here, now. And that's also why we invented time, when we separated from our hearts. Then we are consumed with time because we have guilt, fear, remorse, anger, rage, jealousy, all these things from the past that live with us today. Or fear, which is projection into the future of something that hasn't happened. Every place except now."

Ilarion said that the elders are telling us that we need to be human beings, in the literal sense, because we do not have the luxury of time.

Stan asked Ilarion to speak to the importance of ceremony during this time, particularly since so many ceremonies are dying off for lack of being passed on to future generations.

Elders in Kauai that Ilarion has spoken with say how the ceremonies are "the very foundation for everything. They're more important than people realize. This is why elders throughout the world have been going around the world for the last thirty years to share their sacred teachings."

He went on to speak to how, according to one of the stories he carries, roughly six thousand years ago there was a shift in the balance of Mother Earth into the masculine imbalance that we're still in today. Elders during that time sought a solution to this, given their sacred teachings were feminine-based.

Ilarion compared this occurrence to a piece of paper being torn into small pieces. If the pieces during that time weren't put together, if they weren't practiced for two generations, they would be completely lost to their people.

"They knew that all things feminine are going to be grossly violated: goddess cultures, healers, women, Mother Earth–based cultures and Mother Earth, all of these things that are feminine in character. They were going to be grossly abused and violated, and they knew that these teachings would be taken that way and misused.

"But no culture forgot the same thing as another, so we had this tapestry of pieces around the world. And the only time people would be able to get that back would be if they open up their hearts and trust each other around the world."

During his meeting with Indigenous elders from around the world in Kauai during 2017, they all told Ilarion that they must show the world what they were talking about.

Thus, according to Ilarion, for the last thirty years elders have been traveling the world, sharing what they know, in the prayer that the hoop of the sacred teachings will be made whole once again, and during our lifetime.

"And that means trust, trust in life, trust in your life. Trust in Mother Earth, trust in the universe, trust in that we call 'the Maker,' or the Great Spirit that lives in all things. And that we have no more time to hide these teachings, we've got to share them."

So that is what they did. The film the Reuters journalist helped them release includes showing them in ceremony, and it was distributed internationally. The film has now been viewed by hundreds of thousands of people. Again, the message Ilarion took from that meeting was that there is now just a short time for human beings, and, as he put it, "We no longer have the luxury of trying to connect with the peripheries of those who don't understand. And the ceremonies themselves are very powerful, and they are felt throughout the world with others, but do not focus on that. Focus on just doing your ceremony. Don't worry about how many people it's going to affect."

Ilarion sees this happening now, and cites the shift in how pipelines in the United States have been stopped since what happened at Standing Rock.

"The template was set by elders and they said we must *be* peace. We must be love. And the rest will take care of itself. Don't worry about what's going to happen. And they *did* that, and now we're seeing these changes happen. This is because of the way they approached it. And so that's all I can say about that."

We thanked him for his time, at which point he spoke an Unangan phrase: "Qaaxaalaakux Ungooneesh [Thank you big]," he said. "No meeting happens by accident."

4

Raquel Ramirez
(*Ho-Chunk, Ojibwe, Lenca*)
Awareness

COMPOSED BY STAN RUSHWORTH

> We're all carrying the wounds of our ancestors, and this can be seen in our
> actions and our thoughts.
>
> —Raquel Ramirez

Raquel Ramirez was born on September 11, 2001, and when I told an elder Chiricahua friend of mine this, he said, "With death, there's always a
birth that keeps everything going, and that's a good thing."

She defines herself as an urban Indian, Ho-Chunk and Ojibwe and with
other strong Native family roots, as well as being greatly influenced by California Indian cultures, where she grew up for the majority of her life. Her
family is very much a part of the nationwide Indigenous community.

We meet for our talk via Skype, and Raquel is wearing a bright blue
T-shirt from 1991, commemorating the 125th annual Winnebago celebra-
tion. It was once her father's shirt, and she points out that she has others,
from way back. On the upper part of the shirt, the silhouette of a warrior
rides a spirited horse, holding a lance with feathers trailing in the wind.
Throughout our talk, she speaks with animation and passion and clarity,

even when speaking about the complexities that stymie her, a part of the process, she maintains.

We begin with everything going on now in the midsummer of 2020, and the difficulties many people have putting it all together: climate change, social unrest, the pandemic, history, and how we comport ourselves now. Raquel speaks of awareness as the challenge. "The truth is, awareness is difficult for a lot of people. I think that's why ignorance is so prominent, because breaking the ignorance [in society] is going to be breaking someone's own ignorance, and that's a process, an emotional and difficult process." She says people talk a lot about education, so she defines this adamantly, challenging an oversimplification of what's needed. "If they're truly educating themselves, they will experience emotional and mental deconstruction."

She maintains that this is something many are afraid to do for many reasons, one being because they have no immediate solutions, which is a common silencing by the dominant society and a particular cultural imposition. She brings up the tragedy of COVID on the Navajo reservation. "I had a friend who would argue that if you don't know anything about these communities, or if you can't help these communities, then you just shouldn't know." But Raquel's argument is that "knowing is the first step to helping them."

In this light, she reflects on her part in this book, and on Dahr and my putting it together, speaking about our personal histories of my being a Vietnam-era veteran, and Dahr being a correspondent in the Iraq war. She speaks about PTSD from war zones and times, and the state of the world now, and her desire to contribute "as someone who's interested in the global world." She emphasizes how important talking and listening is in coming to right action. "It's the first thing, the only first thing we truly can do." And, "Thank you for not being afraid of painful ideas and painful experiences," she says, pauses, then goes straight to a core challenge.

"Perhaps our mind-sets have been the problem, perhaps our approach to living on this Earth" is the problem. "I think that allowing ourselves the ability to really change how we see things is how we're going to change things." She reiterates that this is a painful confrontation for many, likening it to dealing with PTSD, and she emphasizes the idea of what we could be "allowing ourselves."

Another perplexing phenomenon she confronts is the escapism she sees. "It really hurts my heart. I don't understand when people say we'll be living on Mars in a hundred years. . . . Are you giving up?! Is that easy? That's solving the problem? Because essentially leaving is easier than changing the way that you think about the world?" She shakes her head and goes on. Again, she presents the fear of confronting how we see.

"What's going on in the environment and in so much of the world is a reflection of personal wounds that have not been healed, for all marginalized communities as well as for nonmarginalized communities. . . . We're all carrying the wounds of our ancestors, and this can be seen in our actions and in our thoughts. . . . Personal healing" in this arena "is our chance to be able to make some lasting change on this world. . . . There's so much pain and there's so much suffering and people choosing to remain ignorant to their own pains, choosing to remain in ignorance, of everything, as a result. . . . We all need to go to therapy as a collective nation." She laughs, then adds emphatically, "Something needs to change!" with no smile at all.

Dahr asks her to talk about her experience growing up as an Indigenous woman in a society bent on "erasure of Indigenous people and Indigenous perspective," including the histories, and how this experience connects to all that she's saying. "It connects well because one of my very first distinct and prominent memories was in kindergarten, and everyone was talking about where we came from, and I brought up that I was Native American. I had a classmate who said, 'You can't be Native American, Native Americans are extinct.' And I think most Native people have had that experience." As I listen to her, I remember my nineteen-year-old son having the same experience, and the fact that Raquel is eighteen years old. I remember Gregg Castro's experience, and he's in his early sixties, so her words reflect a long-standing denial in America. I think of her strong admonitions about changing the "mind-sets," and the choices people make to remain ignorant.

She goes on to describe a childhood in schools where her "cultural practices" were considered "weird," or "not safe," or "dangerous." Reflecting, she brings it to the present. "I think that really extended throughout my life, people not really recognizing what it's like, not really" understanding "Indigenous people and Indigenous perspectives.

"As a result of that . . . I've always felt this visceral feeling that somehow I didn't exist," not so much within family or immediate community but within "the primarily White communities" of school. "This extends to how I see the world . . ." and "there are so many young Indigenous people trying to find their way within this twenty-first-century modern world. They're interacting with their peers, and they're also trying to remain true to their values and who they are. It can be very, very difficult when so many people are telling you that your values are just myths. Really, that's what Indian people did become, a myth. And it's a convenient narrative to tell people because it fits into the story of the United States, of 'We came, we conquered, we've established our land.' And, it's not a clean narrative to say 'Oh, but the Indigenous people are still here today. They're still functioning.'"

Raquel insists that Indigenous peoples "still have ideas about this world. And perhaps they would have something to say about how to take care of the world if you just listened to them. And not just listen," she warns, but develop "awareness. Awareness doesn't just mean listening or hearing or recognizing." It is "very much being present, and being conscious of people beyond you!" This is the key, "beyond you," the cultural value she holds to. As she articulates her thoughts, she struggles with the word "awareness," saying she's "trying to express something greater than that," and this becomes a central thought she develops.

She talks about herself as an "urban Native functioning within White spaces," asking for something much more from her surroundings than is being given, expressing how it all "can make you feel diminutive, very, very, small." But she says, "I've spent the past year being able to try to heal from that, and learning to heal, and learning to love myself." She takes that work out into what she sees as her larger task, "to best serve my communities." Again, it extends to being "conscious of people beyond you," an admonition to herself that she wants the surrounding world to imbibe.

Within this process are emotions she says she needs to describe, complexities when dealing with the meetings between Indigenous and non-Indigenous people. "It can be very frustrating when a White person will come and ask for Indigenous wisdom that can solve climate change. There's a certain kind of conversation out there that really just disregards how you're asking people to just keep on giving you more and more, and you're not taking into account that they've given so much." She pauses, looks at the ceiling, then back.

"This, without allowing Indigenous people the ability to grow and heal. It's something I get frustrated about. It's that, 'Okay, you want us. Now you want our voices, but you don't want us to heal from everything that's happened.' It's a very selective process to say, 'I want the Indigenous wisdom, but I don't really want to hear about your pain and having to adjust to what we have created and continue to create for you.'"

She says this is why thinking deeply about all of it is so important. "If you can listen to the pain of Indigenous people, you'll be better able to understand the pain of Mother Earth!" "The pain we're carrying" is not only "the pain of our ancestors and the pain of genocide. We're carrying the pain of our home being destroyed. And now we are expected to try to save the world?!"

From cultural collision, Raquel then talks about the "cultural fusion" of her upbringing. "I am Ho-Chunk and Ojibwe, but I grew up with a lot of California Indian practices." She describes "urban hubs, Native hubs all coming together to create a space. . . . We have to recognize the individuality and uniqueness of each tribe, but I have also found it very important that we find our unity with one another, our connection with one another," and "urban spaces are great places where that happens, where I hear Native voices from all across the state, all across the country." Her grandmother, Dr. Renya K. Ramirez, writes extensively about these connections in her book *Native Hubs: Culture, Community, and Belonging in Silicon Valley and Beyond*, and Raquel's life and words reflect the complexity of the landscape her grandmother explores.

We talk further about the sense of "cultural fusion" between Native cultures, now also alongside the question of including non-Native people, specifically in ceremonies at this point. The question of people wanting "Indigenous wisdom" brings this up, in the tone of entitlement she's mentioned. "There's a history of White people just taking, but I also don't like the concept of exclusivity. Again, it's difficult. . . . There is a certain value in the exclusivity of protecting our spaces, and not just because of the history of what's happened, but also in that these spaces are safe spaces. I know that's a popular word, but it does mean something." She describes "irreverence and disrespect" she's seen from non-Native people in Native spaces, and how this can make one feel "on the defensive. . . . A lot of Native people are going there for their own individual and personal healing, and it can be disruptive. Sometimes I don't think it's our duty as Indigenous people to have to be open and welcoming. I know that's kind of contradictory, but it's not our job. . . ." At the same time, she addresses another side. "I have complex thoughts about this, but I do think that we need to promote community and connectedness amongst all people. This is first and foremost a mind-set that we approach the world with."

Everything she says about this echoes back to her opening comments about "the ability to really change how we see things." People cultivating this ability may allow the fusion she promotes on a broader scale.

After speaking briefly about the long history of taking life and land from Native peoples, Dahr asks, "Why is it important for us all to go back and really acknowledge what was done to Indigenous populations on this continent? Why is it important to go to the root?" Raquel thinks for a moment, but without hesitating, she elaborates her thoughts.

"Paramount in Indigenous ideology is the nature of cycles and the circle, like our medicine wheel. There is a constant cycle in human experience of struggle and growth. And until you can experience struggle and you can experience pain . . ." She pauses, indicating that without that emotional "allowing," it goes nowhere, then she explains her point further. "We experience the most amount of growth if we allow ourselves to reflect, and we can start reconstructing. For generations of Europeans not to reflect and reconstruct, they're not participating in the cycle of human experience." She maintains that their not reflecting "breaks the cycle."

"The genocide hurts the heart of the perpetrators as well, because it is such a stark contrast to who we are as human beings, inherently collaborative, inherently community focused. And for you to kill people, that's a wound on your heart. And so they have that generational wound as well! They carry the generational wound of genocide as well. And by not reflecting on that, and by not growing from that, they're stagnant." She stands on the last word, pauses, then continues.

"That's why it's so necessary for there to be *awareness* [this word returns and grows], because it allows human beings to work within the cycle of Mother Nature. . . . That's how Mother Nature works. We have things that grow with things that get destroyed and the things that grow from ashes. . . . We have to be reflective in how we view the world. We grow, we struggle, and we'll have to grow some more."

Throughout her talk, I don't see Raquel become angry, but here she shows deep annoyance and impatience. "And this is why nihilism is so frustrating for me as an Indigenous person. 'Oh, the world's over, let's leave.' You are not allowing yourself the opportunity to grow and to feel this pain! Nihilism is easier . . . easier than hope. . . . So that's why we have to be aware, because it's a broken cycle!

"This lack of awareness is the reason why we are destroying Mother Earth. We don't see our place in the world, our place within the system." She says people are caught in the idea "of humans being the superior species." But, "Our purpose is to function within this space because we are of this Earth, and so not being aware of what our place really is, has resulted in us being able to disrespect, to desecrate, to harm the land."

She quickly moves to the necessity of "having a sense of responsibility" and quotes Dr. Darryl Wilson: "There's an awesome power that binds us all together." She goes on, "My actions are not independent to me. We're obsessed with individualism and independence, but we're not independent of one another." She describes how colonization and the "narrative of the pioneer" and the current economic model is about "conquering" people and land. "And if you can't be aware of that, you're going to continue to exacerbate those practices." Again, she asks for a hard analysis of perceptions, the legacy we live under, the influence on our "mind-sets," and the courage to do that work.

She talks about what gets in the way of that analysis, what stifles the courage, which she says she thinks a lot about, with many people asking her how she can deal with the hard things in her own life as well, the tragedies. "I don't really know the answer, and perhaps it's my Indigenous background that allows me to reflect on personal traumas." She maintains that not reflecting on the hard things in life "was never really an option. . . . Maybe from a really young age it was ingrained in me to really notice. My ancestors and I had to recognize our past, and I had to learn from them. And I had to honor them, and that gave me the mind-set that when difficulty happens to me, I continue the same practice of analyzing and learning from my past."

So what gets in the way for so many people, too many people? "A lot of it's just fear." And, "It's a reflection of people's inability to reflect. . . . Anybody who's suffered from PTSD or any sort of mental strain can know that the process is grueling and it's uncomfortable. . . . We are so obsessed with comfort and ease that just sitting with it and acknowledging it 'hurts' and is 'painful.' So it can be very difficult for a lot of people," but she emphasizes again that it's necessary. She softens, mentions the need to feel compassion, then adds that some people feel that facing trauma is letting it run their lives, that they're "letting it get to them." But that's not her way.

She insists that "people can acknowledge what has happened in this country, what has happened in countries all over the world where Indigenous peoples live," then she further addresses this major impediment or refusal that must be faced. "There are some who think that admitting this is a weakness. So they choose to just ignore it because they value strength as not being vulnerable. Being invulnerable is considered strength." She explains this with some exasperation at the idea, but she explains that "vulnerability is the ultimate strength, and being vulnerable in your personal life extends to being vulnerable in the broader context. . . . It's vulnerability to acknowledge how small you are," yet "how much you impact everybody. It takes a lot of vulnerability to be able to acknowledge everything in this climate crisis, and in this social and political crisis. . . . I used to think it was an option. You could ignore it. But I don't think it's possible to ignore it any longer. And it's just adding another layer of denial when folks pretend to ignore it, so this then exacerbates the problem."

With all of her points in mind, we begin to talk about the way forward. Raquel begins with recalling Ella Cara Deloria's novel *Waterlily*, a book about early-contact Indigenous life that focuses on child rearing. "If we really want to start changing the way that people view life, and change people's ability to embrace difficult experiences and be aware of them, it needs to start with the children. It needs to start with the parents. It needs to start with people learning from a very early age how to handle these things, knowing that they're safe when they're vulnerable, knowing that they are safe to be themselves, from a psychological perspective.

"Mentally, we have a lot of sickness, a lot of mental illnesses. A lot of selfishness. If we can start with our children, we can start with the future. And we can all start individually. We can start the process of healing and growing. . . . And this will extend out into the world. If we really want to get down to it, we have to start changing how we treat our children. If we can treat them well, they'll be more compassionate to themselves and to other people, and they'll know they have responsibilities. This will extend to their becoming functioning human beings." As she speaks with emphasis on the last two words, I think about how many Indigenous tribal names translate into "human beings."

Dahr asks what she sees herself doing in the future, as part of the way forward for all people, how she might realize her life goals within the current situation. "I do not know the future. I cannot predict the future, so I feel less and less comfortable saying 'This is what I want to do.'" She is clear about one thing, however: "I want to be of service in some capacity to Native people." She speaks clearly of "being of service to the community, and also being of service to the people that I really love. But, I don't know how to market that on Craigslist." She laughs.

She describes past experiences about her education: "I would struggle with motivation with school because I was feeding myself with this idea that I needed to function within this White system. I needed to prove myself within this White system, and it was demoralizing and disheartening because I did not find those things motivating." But she changed, again through reflection and seeing the bigger picture, through gaining the courage to look at it all. "I started to change my mind-set by really focusing on my studies and learning to just love learning." And now, she puts it all together. "I'm becoming a kinder, more empathetic, more understanding person. The more you're willing to educate yourself, I think the kinder you will be. What I really want to be is a kind person with a positive impact on the world, in any capacity."

Dahr asks if this has to do with the difference between being raised with "obligations" rather than "rights," which Raquel had brought to mind earlier. Dahr offers the question as a possible contrast between Indigenous and non-Indigenous approaches to life in community in the largest sense.

"It's interesting," she replies. "I think my rights stem from my obligations, versus a colonial mind-set where your rights free you from obligations. You know, being a human on this Earth means that I have obligations. It also means I have rights, but one comes first. If I can fulfill my obligations and my responsibilities, from there my rights come, my rights of freedom and my rights of finding happiness. . . . These words have negative connotations, but no, I should want, as a human being, to be obligated to my family and be obligated to my loved ones." In her cosmology, this is a very large family, and the size is apparent in her gestures as she spreads her hands wide in the air.

She then addresses human obligations within collapsing systems, both now and in the near future. "There was a tweet that went viral, that was about how Wow! You're having your world around you crumbling and you're feeling like being invaded by others. This is a new feeling for everybody. As an Indigenous person, I've seen this on social media recently," and she adds that this shows her that "there's more and more awareness that White people are experiencing an awareness of all the failings within the system." She hesitates, then adds, as a Native woman, "But we've been feeling like we've been on the decline since first contact, like the decline hasn't just begun, it's been ongoing.

"But the key difference is the awareness, right?" She returns to her initial idea. "And if you want concrete proof that awareness does something, look at what's going on right now!" She refers to everything going on, very literally, awareness and lack of awareness around the world and on Turtle Island.

She makes a point to say that she is not saying "I told you so," because she's "not going to criticize someone for spending their life not knowing." She says she understands people feeling deep frustration and fear, from the "awareness of all the failings," but she emphasizes that for her, "it's okay to be late to the game. . . . For me it's just that. So okay then, 'Let's get started.'" She repeats, "Let's get started!" She smiles, then returns to her idea of stagnation.

"I'm glad that people can see what's going on. They're healing and they're getting out of a place of emotional numbness. I'm grateful that we're removing those shackles, and that people are changing.

"It reminds me that the . . . *obligation*," she starts to say, then reflects that "it's not the right word." She searches for it, then says, "The word is love. That's what I mean when I say that my rights derive from my obligations. Really, my rights come from my love and my capacity to love others and love this planet and love individuals and love communities and people that I haven't met.

"This is what gives me my rights as a human being. This is what gives me my freedom. And paramount is the freedom to love, by learning to conduct ourselves with love, by loving ourselves and loving other people. The importance of love is not really recognized, and destroying the planet you live on is showing an absence of love!

"It's difficult to bring it all together, but that's the thread of what I've been trying to touch on. We have to love our children better and love ourselves better. We have to love our communities better and we have to love this Earth better . . . and we have to learn how to do it . . . and we have to *want* to do it.

"So maybe if we can change this idea that it's going to be difficult or it's going to be hard," and know "that we shouldn't run away from difficulties, but we can transform them into this idea that if we can learn to love better, this is how we will be changing the Earth. That's how we can change the Earth.

"It sounds so vague and nebulous, but it's very, very concrete. Like my mom would say, 'Well, can you implement it into a policy?' Yes, you could!"

I know what Raquel says is true, and I can see her forming laws from her relationship to Earth, to all living things. I can see this in our future, through her words and surety.

Finally, I have to ask her about her own visceral relationship to Earth, and she reflects, then smiles. "As a child, I'd be in the sweat lodge and it'd be really, really hot. I was four years old and I couldn't handle it. And my dad would always tell me to put my face on Mother Earth, and that she'd calm me down. It is one of the most powerful memories I have, of putting my face on Mother Earth. And it's receiving that love and being open to that love. That's what so many ceremonies are based on, just like finding love within yourself." This experience is the origin of the love coming from Earth, and her love for Earth, for self and all others, and of her awareness of this as her obligation.

"It's literally a metaphor, where the world around you is so hot and suffocating, and, you know, you put your face on Mother Earth, and she will protect you."

In closing, I can think of no better advice, and I remember all those elders who've presented this fact in one way or another, and now I'm with a young woman who knows this and speaks it into the world, a great blessing.

We thank Raquel for the talk and her insights, and she replies, "Well, this is not me. This is everyone. When I was speaking to you, I was thinking of my grandmother and my uncles and my aunts, my grandparents, my great-grandparents, and my friends; this is an amalgamation. I am multiple pieces. And I'm an individual piece, and, I've just listened to my elders."

We have just listened to her, and we are grateful for it.

5

Lyla June Johnston (*Diné [Navajo], Tsétsêhéstâhese [Cheyenne]*)

Trust

COMPOSED BY DAHR JAMAIL

It is through the release of fear, and the trust in Creation, trust in ourselves, trust in our inherent ability to be taken care of by this Earth, only then will we be free.

—Lyla June Johnston

Lyla June has a degree in Environmental Anthropology (with honors) from Stanford University and a degree in American Indian Education (with distinction) from the University of New Mexico. At the time of this publication, she is a PhD candidate in the University of Alaska's Fairbanks Indigenous Studies Program, with a focus on Indigenous land stewardship. Weaving this knowledge into her speaking and multigenre art (prayer, hip-hop, poetry, acoustic music, and speech), she has inspired international audiences toward personal, collective, and ecological healing.

Having grown up with an Indigenous worldview, coupled with her education, sets Lyla June in a place where she focuses her messages on Indigenous rights, supporting youth, intercultural healing, historical trauma, and traditional land stewardship practices. Her stated personal goal is to "grow closer to Creator by learning how to love deeper."

Lyla June is a busy woman. Scheduling a longer interview with her took months, as her service work leaves her with little spare time. We caught up with her after she had finished meeting with elders to produce an online gathering for hundreds of women.

Stan began by asking her if she would speak, generally, of the time we are in, of the converging crises of climate disruption, ecological harm, social instability, and the global pandemic. Lyla June took a deep breath, collected her thoughts, then began.

"It's understood by Diné people that we've been through the destruction and rebirth of several worlds before this world we're currently in," she said, speaking very slowly and deliberately. "One of which was destroyed by a flood. You see that story replete throughout cultures across the globe, which leads me to believe it's not just a myth."

"Each time that the people came to an impasse," she continued, "they faced a decision to evolve and change, or perish. Now, on Turtle Island, we have experienced a lot of challenges prior to Columbus's arrival. We actually have experienced epidemics, collapse of social and ecological systems, slavery, caste systems, warfare and discord before Columbus ever got here."

Lyla June went on to explain that one of the places this happened was Chaco Canyon, in New Mexico. While many archaeologists hail the place as an archaeological treasure, the Diné never go back.

"We view that place as a place where we fell off the path very badly. We had a caste system there. We tried to play God there. Negative spirit was working through us there. And at a certain point, the Diyin Diné, the Holy People, they sent us a drought and this drought destroyed our civilization." She said the drought was "a great gift to our people. It was the only thing that would give us the courage to change."

Lyla June has heard it said that it was the youth who rose up during that time and said, "No more. No more of this hierarchy. No more of this supremacism."

Her people abandoned that place of suffering and pain, and they never looked back. She pointed out how the same is true with some Mayan civilizations. But while White archaeologists are arguing over why these civilizations collapsed, they never asked the people. All three of us laughed at that.

"It didn't just collapse, we left that place, that place where we tried to play God, that place where we fell off the path," she said even more slowly. One of her elders asked her, "Can you imagine the faith that it would take to leave your city and go back to the forest?"

Lyla June paused there, then said, as a matter of fact, "So we've experienced all of these things before. The beauty of our languages, the beauty of our cultures, was born of ashes. We earned all of this wisdom the hard way. We've had warfare. You think about the Haudenosaunee and their Great Law of Peace and the Peacemaker who came to help them. Why do you think he came? There was so much war."

Lyla June, being an anthropologist herself, believes the depiction of Indigenous peoples portrayed to us by archaeologists and anthropologists is sorely distorted and inaccurate.

"We densely populated the land. We managed it extensively, through everything from seasonal burns, to thinning of the forest, to propagation of edible plant species, to the maintenance of whole watersheds. We managed the land on regional scales."

Lyla June said her people learned all these things by way of trial and error, which is one of the key tenets of Diné fundamental law. "As my elder Philmer Blue House says, we have four databases within Diné fundamental law. One of them is negative wisdom and positive wisdom. Negative wisdom, you gain through pain and challenge. Positive wisdom through feeling what works. Negative wisdom is what we're going through now."

Lyla June went on to point out how societies have risen and collapsed in isolated cases, all across the globe, not only here. "For thousands and thousands and thousands, hundreds of thousands of years. But now we're experiencing collapse on a unified global scale. And from those ashes, something will be reborn."

She believes and emphasizes again how important it is that we understand that something like a drought is not a curse, but can instead be a gift. "[It is] the only thing that will help us change. The only thing that will wake us up," she said, before going on to address exactly what I'd experienced with over a decade reporting on the climate crisis by way of presenting facts to people, hoping to inspire change. My book on the climate crisis had twenty-one pages of footnotes, which I'd naively hoped would awaken the broader public to the gravity of the crisis upon us. Alas, most of the general public, at least in the United States, still behaves as though there is still time to avert the worst of the crisis.

"We can have all the science, and all the charts, and all the parts per million, all the atmospheric projections in the world, but apparently that doesn't inspire us to change," Lyla June continued. "And it was prophesied long ago that when we arise from this, generations later, *as was prophesied long ago*, that this world would end and a new one would begin if we did not heed the messages of the Earth and the messages of the ancestors. It would be reborn, and that's happened many times before."

Lyla June went on to explain how in the Diné understanding, we've already been through four specific worlds, and Coyote was there at the beginning. "Coyote is our brother, but he's really unhelpful," she said, then laughed. "He's always tricking and trying to gain power and hurting people. And that's a real spirit. That's not a story. He works through all of us. We give up our power through fear. And it was fear that got us here."

Hence, she sees this dilemma of fear as akin to "Chinese handcuffs," in that the more you struggle to free yourself from the fear, the more entrenched in it you become. The solution? It is through the release of fear, and the trust in Creation, trust in ourselves, trust in our inherent ability to be taken care of by this Earth, only then will we be free." Until then, the way Lyla June sees it, many of the current solutions being proposed are actually part of the problem.

"You can't win war with war. You can't win fear with fear. But they're going to try. It was fear that caused men to hoard, and it was hoarding that caused our planet to die." Again she comes back to the solution, one known to her people who have already lived through collapse.

"So [it is] only when we trust that there is enough. When we find mechanisms like Coyote stories or prayer, ceremony, then we can release fear from our societies, the way that our societies learned to do after the collapse."

Her people believe that Coyote continued to try to get into their societies after that, but, as she said, "We had all of these little mechanisms, understandings, even built into our language that prevented it from taking hold." But it is different for the society of the current dominant culture.

"In this society, we have the opposite. We have a lot of mechanisms and things in our language and social norms that make a *lot* of space for Coyote to thrive. We're learning through his tricking. He's tricking us really bad right now, and it hurts. It hurts our women, it hurts the water, it hurts the men, it hurts the children."

She believes we have to find the mechanisms that are allowing Coyote energy and the fear that comes with it, and change them. "We have to re-create those mechanisms again, and we *are* doing that. We are. Slowly but surely, we are."

Lyla June understands, deeply, that we are in crisis, and returns again to the essential need to transform the fear, to see the situation as "our gift," difficult as that may be.

"It is a gift to humanity to give us the courage to change so that we can hand something to our children that they can be proud of. They can be proud looking back at what their ancestors did. That we walked away from the city and we went to the forest. We realized that the forest is already a city, and that we can manage it in good ways."

Lyla June, prepandemic, traveled domestically and internationally doing her work, and has met with Indigenous people all along the way. Stan asked her about responses she is seeing as she has done her work. She cited the Black Lives Matter movement as an example of how we are rejecting hierarchy, and the youth climate strike reminded her of the youth who rejected what was happening in Chaco Canyon.

"They are saying, 'No, no! We're not going to play this game, and we're going to grab you by the collar and say, "We see you."'" There is an army of Indigenous people who are speaking about taking care of the land in our traditional scientific ways, and there is the fact that people are not just listening, but searching for that Indigenous wisdom. We've already gained a good amount of negative wisdom by now. We've already seen that this doesn't work." She sees the aforementioned, as well as the example of Occupy Wall Street, as humanity's rejection of the cultural norm of capitalism, as indications of our learning, as well as assertions that people are seeing through the lies of Coyote, and rejecting those lies.

We asked Lyla June if she would speak about unresolved intergenerational trauma, as well as her use of the word "holocaust," as opposed to the word "genocide," when she discusses what was done to Indigenous people on Turtle Island. She began by talking about the fact that when European colonists came to Turtle Island, they saw incredibly advanced societies.

"They saw the Haudenosaunee with beautiful, complex, and efficient systems of governance. Matriarchy was woven into all of their practices. All the explorers write in their diaries that the land looked like a vast park, and that's because our lands were taken care of. Pruned and managed in a way that fed not just us, but all of our relatives, human and not. We populated the land in dense numbers, and at the time, these European colonists had undergone severe trauma for thousands of years." Lyla June paused there, before going into greater detail.

"It was everything from the Roman expansion, to persecution of women as witches, to the destruction of forest cultures, which they called savages. In French, 'savage' comes from the word for forest dwellers. And they destroyed the forest dwellers of Europe before they destroyed the Native people of Turtle Island. The Inquisition, King Edward the First ordering the massacre of a hundred Celtic harpists, the British kingdom destroying the entire oak ecosystem of the British Isles that had been maintained by people there for thousands of years. They would go into the Scottish communities and they would find the bards, the storytellers. The bards carried the songs, and they would kill them first and bury them face down so their stories would die with them. They would force every woman who was married to sleep with the British lord before she could marry her husband." Her voice tone grew lower as she told these stories, the heaviness of them tangible as she spoke.

"Horrible things. The prohibition of Indigenous European languages up until the 1920s; the Welsh language was prohibited in schools. If you got caught speaking Welsh you got a block of wood tied around your neck with the letters WN written on it, which stood for Welsh Not, and the only way you could get the Welsh Not off your neck is if you caught another kid speaking Welsh. Just thousands of years of intense suffering, torture chambers, and the destruction of the human spirit."

Lyla June believes Coyote had a lot of room in that world, and essentially made that world. "His little kingdom," she said. "Some kingdom, where everyone's hurting."

So when the colonists came to Turtle Island, Lyla June sees them as a people who had something similar to narcissistic personality disorder, where their self-esteem and sense of self was so vacant that whenever they saw something greater than them, they had to destroy it. "Their self-esteem was so low that when they came here and saw our civilizations flourishing and doing well, and that the women were taken care of, they hated that because they came from a land where *they* were supposed to be kings, *they* were supposed to be the best. They were the peak of civilization in their minds, just like a narcissist thinks that they're the greatest person in the world."

Lyla June paused, then continued speaking matter-of-factly. "And so they destroyed us," she said. "I think White supremacy is akin to White insecurity." She laughed at the irony. "Very insecure. Why else would you go around destroying people to make yourself elevated? Unless you didn't have a secure sense of self to begin with, and that was Coyote's doing over many thousands of years."

She then took us further back in time again. "Constantine was the first Roman emperor to begin conquering people in the name of Christ. This was in the years around 300 A.D. So they had actually been appropriating

Christ for conquest for over a thousand years before they came here and
started doing that to us. That's cultural appropriation at its most perverse
level. To take Christ and his teachings and his love and distort it, twist it,
destroy it, and then use it to justify the slaughter of other people. That's all
a part of the mechanism of White insecurity, to say 'My religion is so high
and so holy and so great, and so wonderful that if you don't have it you
aren't even human, and I have a right to destroy you in the name of my
religion.' And that happened."

Stan and I both felt the weight of her words, as the history of the violence
moved in the air around us.

"We were enslaved in the name of Christ. We were destroyed in the
name of Christ." Lyla June took a deep breath, then released it, before continuing. "So when you look at the continent of Turtle Island and what
happened here, not just in 1492, but even prior to that, with the Norse who
came, they who liked to fish over here, according to my elders. They liked
to fish in the northeastern part of this continent, and when they came they
would steal the women, and this made us afraid, and so we moved inland,
and the fear of them gave us disease. Fear is the root of disease. And we
died in great numbers before Columbus ever got here."

Lyla June went on to say that the idea that this continent was isolated
prior to Columbus's arrival is false, but the situation began to decline dra-
matically after 1492, each century becoming worse than the previous, and,
as she said, "Especially the eighteen hundreds, things got really bad here.
Really, really, really bad." She paused as she was crying, to wipe away tears.
"Really bad. Coyote was just having a ball destroying us. And there's a reason Coyote targeted us: *because* we were close to Creator. He expended a
lot of energy. Because Coyote has limited resources too. He chooses his
battles. He expended a lot of energy on us."

Lyla June said the Europeans became the puppet for all of that destruction, and rather than choosing to be good to other people and other things,
allowed their fear (in the form of insecurity, coupled with the fear of un-
worthiness, and other fears) to take them over. "And as soon as you have

fear, you give Coyote power. You become a vessel for that, but it's all an illusion. None of that power is real. Coyote has no power, only the power we give to it." But Lyla June believes we are taking that power back, by way of naming it and giving language to it: intergenerational trauma, racism, colonization, scientific racism, supremacism. She sees this as the start of a rising out of those ashes.

Then she went on to talk more specifically about Indigenous people on Turtle Island. "We need to protect Indigenous peoples from this ongoing genocide," she said, before going on to further explain her use of the other word to describe what has been, and is being carried out against the Native peoples here. "We call it the Native American Holocaust because we want to show people that everything we abhor and judged in Nazi Germany is going on right here. The deliberate destruction of an entire race of people."

Lyla June sees that as changing a little bit now, and acknowledged that people are more sympathetic to Indigenous life itself, but reminded us how in the 1800s, the policy of the U.S. government was literally that of extermination, which was written into law, is extremely well documented, and included state-sanctioned bounty hunting of Native peoples.

"The last thing I'll say on that is that the policy of extermination changed in large part due to an historic event known as the Sand Creek Massacre," she said, before adding her feeling proud to have some Cheyenne ancestry. "They turned in their weapons. They surrendered at Sand Creek. They told the government that they were done fighting. They said 'We don't want to fight. You are trying to turn us into a fighter, but we don't want to be a fighter. We don't want violence. We don't want guns. We are people of peace. We're not gonna fight you anymore. We're not going to do it.' And they formally 'surrendered,' but perhaps it was more of an assertion that they are people of peace."

After taking a deep breath, and a moment of quiet, she continued. "The U.S. cavalry came in *knowing* that they had done it for peace, *knowing* that they gave up all weapons, *knowing* that that was their conviction. They came in and slaughtered the people: men, women, children, elderly. Murdered them. Hundreds of them. And then after they murdered them, they mutilated their bodies because the Cheyenne thought they couldn't cross over to the spirit world unless their body was whole." She wept as she told the story. "Soldiers *knew* that, so they chopped off limbs, did very bizarre, horrific things to our bodies. There are even stories that they turned our men's scrotums into bags. There is a book at the University of Denver which has a cover made of Indigenous people's skin, and they didn't take that book down until the 1970s.

"And so the way that people carried themselves that day, the way the Cheyenne people walked the Earth that day, it shook the whole world, and America could see plainly who was acting savage. It shook the core of the colonizing culture, and extermination was no longer acceptable, even to the colonizer. It rocked the whole country, and it changed American policy forever."

Today when people tell Lyla June that nonviolence doesn't work, she points to that story. "My elders told me that, ironically, even though much Cheyenne blood was shed that day, the way the people carried themselves is the reason there's even one drop of Cheyenne blood on the planet today," she said.

Then the change in policy occurred. "So they switched to assimilation. They went from extermination to assimilation, which also sucks," she laughs. "But at least we're still here. Right? I'm pretty sure that's the reason I'm alive today. It's because of the way the Cheyenne walked."

Stan talked briefly about our hope of this book serving to help alleviate the fear that is so prevalent now amidst the converging crises, and mentioned how his elder, Darryl Babe Wilson of the Iss/Aw'te (Pit River Nation), had also spoken often of Coyote. He asked if Lyla June would speak about courage, and how we might find it, in the face of so much fear.

"I think your intuitive desire to have this book cut through the fear is a good reminder to me, and to everyone. Spoken like a true elder, Stan. We need to let go of the fear." She paused, taking some moments to reflect further upon her answer.

After taking a deep breath, Lyla June decided to speak to her healing process as a way to answer the question.

"It is hard to have true courage unless we take the time to shed the boxes that this world puts us in and arise as what we are. From what I understand, this is done by reviewing our lives and rejecting all the ways our childhoods and lives taught us something wrong about ourselves. And that is something each and every person has to address within themselves. It's a microcosm of the larger phenomenon of intergenerational trauma. And it is through that healing of the self that we see through the lies and have deep courage and trust in the goodness of our being. I am not there yet, but this is what I have seen some others do.

"Let me give a concrete example," Lyla June continued. "Let's say you're abused as a child, sexually abused perhaps. And you grow up thinking that this episode in your childhood makes you dirty or tainted or ugly, or somehow fundamentally broken. Then you don't have a stable foundation to have courage in the rest of your life because every time you stand up, you think people can see through you, see that you're corrupted somewhere deep down. That happened to me, and it happens to a lot of people. And so until you go back there, which I'm in the process of right now, and heal the misunderstanding. . . . A lot of healing is just correcting misunderstanding and saying, 'You know what, just because something bad happened to me doesn't mean I'm bad, and I'm not going to blame myself for that anymore, and I'm gonna put the blame where it belongs, on the person who was used by Coyote to hurt me, and I'm going to feel it and look at it in all of its cruelty, all of its selfishness and heartlessness, and I'm going to feel that. And then I'm going to forgive that, and I'm going to let that go because that doesn't have anything to do with me, really.'"

She paused, then added, "So doing that when you have so much sexual abuse in your life, you have to really work at it. There's just so much, it's like a mountain to overcome." After another pause to reflect, she repeated what she'd said earlier: "So I think that part of having courage is healing."

We were both deeply moved by her openness, courage, and deep sharing. Stan thanked her, agreeing with the truth that her talking about healing the self helps to heal the broader world, and he spoke briefly about how they go back and forth with each other in that way. He shared with Lyla June his experience of his students, especially his non-Native students who are carrying the transgenerational trauma from being descendants of the abuser, due to the genocide not being named.

The time we had with Lyla June was nearing an end, so we asked if she would speak about her perspective of the core values of her Indigenous culture.

"I always refer to my elders because I feel like they have the only things worth saying. They talk about how long ago, and still today, we also had competition in our societies. But it was a competition of who could be the most humble, who could be the most kind, who could be the most loving, who could be the most generous, and who could be the most courageous, not in an egotistical sense, but in a communal sense."

She spoke of how it was different back then, a friendly competition. Those were the goals for the people, and they nonetheless were all learned through having very harsh experiences with the opposite of the positive things. "But we got to a point where that was our goal," she said. "And so you see in different nations, like the Heiltsuk Nation of Bella Bella, British Columbia, they have their big potlatches, and the most venerated person is the one who gives away so much."

Lyla June concluded by speaking to the importance of humility. "In Diné culture, the word 'ajoobá' means humility. That was the guiding star, the North Star of the people. And so when we try to walk that way, we bring health and wealth to everyone."

We thanked her, and expressed deep gratitude for what she brings to the world, and for the generosity of her spirit. Lyla June thanked us, and said she would be praying for us to have all the support, inspiration, safety, and energy we needed to complete this book.

"I know it's a huge task, so just continue to follow your heart and put one foot in front of the other," she said. "And I know that Creator can help you complete it in a good way."

6

Dr. Kyle Powys Whyte (*Potawatomi*)
Kinship

COMPOSED BY STAN RUSHWORTH

Kinship refers to relationships of mutual responsibility, where we care for each other, and we create bonds with each other that make it so that, regardless of what the law says, and regardless of how severe a problem is, or regardless of what our rights are, we have an abiding sense that we need to care for others.

—Dr. Kyle Powys Whyte

I first met Dr. Kyle Powys Whyte, professor of environment and sustainability at the University of Michigan, at a presentation hosted by the Feminist Studies department at the University of California, Santa Cruz. I was quickly moved by the importance of his message, as well as by the immediacy of his presence and the manner of his engagement. This was borne out by the comment of my son, a high school student at the time: "I like this guy. He's telling the truth, and I get it. I don't understand every single word, but I get him." I say this because there was a clear kinship felt by my son, on a complex topic, which sheds light on Kyle's thoughts.

Going into the interview, Dahr and I were a bit unfair to Kyle in that we told him we wanted to give him the lead, but then we asked him to address specific things as well, the first being the question of how we got here to this climate crisis and what's connected to it. We brought up the sense of panic in societies around the world, and asked him to talk about how that panic might be differentiated from a sense of urgency. Asking a philosopher to limit himself to certain terms is a bit of a stretch, even an imposition, but he was gracious in his response. Finally, we asked for his thoughts on how, as members of the larger world society, we might comport ourselves as we meet the situations before us all. He nodded, and dove straight in.

"There's the menu. Sounds good. I can absolutely work on the question of 'How did we get here?' There are a lot of ways to understand what the 'here' is in terms of what we've gotten to, so I'm going to focus on one angle.

"With issues like climate change, and with the public health crisis, there are problems and pending crises for the future, but seemingly both the problems and the solutions are ones that people can't seem to address without harming each other, without committing violence.

"Climate change is an example of this. We see all the different climate change solutions to create renewable energy, but the literature is growing that even these solutions are ones that are bad for Indigenous people. They're bad for people of color. And it makes one think whether a future with some of these solutions is actually going to be any better for Indigenous people and other groups. And so we're in a time where I'm pretty profoundly skeptical when people are talking in the solution space, about whether they're actually valuing groups of people who are disempowered today, and who are vulnerable to being harmed even by the solutions to some of the leading problems.

"In terms of 'How did we get to this point?,' when I look at some of the problems that we see in climate change and public health, there are some things that really stick out to me. One is that people are failing to treat each other with a commitment to reciprocity. They're failing to value each other's consent. They seemingly don't take seriously the importance of trust. There's a sense in which both transparency, when it's appropriate, but also privacy, when it's appropriate, are not considered serious values. Obviously Indigenous people are constantly facing issues of privacy regarding the use and abuse of Indigenous knowledge within public and corporate and other spheres in cases where it's not supposed to be, and doesn't belong.

"If you look at reciprocity, trust, consent, transparency, privacy, or confidentiality, these are all what I call the qualities of relationships we have with each other. And they're a key aspect to what I would call a part of kinship. Kinship is a certain type of relationship that we all have with each other. No matter what society you live in, there's going to be some degree of kinship. It just might look different or be exercised differently, or it could even be lacking, or particularly strong in some societies as opposed to others. But kinship refers to relationships of mutual responsibility, where we care for each other, and we create bonds with each other that make it so that, regardless of what the law says, and regardless of how severe a problem is, or regardless of what our rights are, we have an abiding sense that we need to care for others.

"We need to be responsible for each other, and that's not just confined to the human to human context, but depending on the culture, to all living beings and nonliving entities and systems. So responsibility is a really important way of understanding kinship. Responsibility is a type of relationship. As we know, because we have a responsibility in relation to somebody else, that's a shared responsibility, but it's only going to work if each of us has a profound respect for reciprocity, for consent, for trust, for transparency and confidentiality. You can only really have a responsibility that's working and functioning if you have these qualities. A lot of Indigenous scholars and knowledge keepers, or knowledge gifters, talk about reciprocity or about consent or trust.

"What I would argue is that the position we're in today is one in which there is a profound lack of kinship, and the decline of kinship is traceable to the impacts of a very complex web of systems having to do with colonialism and capitalism, industrialization, ableism, and patriarchy, and over time, these different systems have had different expressions.

"Locally and on a global context, they have disrespected, erased, and continue to ignore the importance of kinship. One of the things I reflect on is just the harms and atrocities of colonialism. For example, oftentimes folks will point to the idea that Indigenous people historically also did some things that were pretty bad. I think about historical cases of Indigenous people. What I notice is that even though at any point in time, any society will have issues and problems, with Indigenous societies, Indigenous traditions of memory making still acknowledge kinship as a way to interpret the causes of and lessons from cases of badness in the past. If I had to bet who would be a more peaceful society, I would argue that a society that would not forsake and ignore kinship would have a better chance to support people's increasing freedom and safety and well-being than a society that just sort of throws kinship into the dustbin and doesn't really care for it.

"When you look at problems that exist in the world today, with climate change, oftentimes the reason for further problems is that creators and builders of projects don't care about consent. There have been numerous cases of where a wind power project or a hydropower project is extremely harmful to Indigenous people.

"If you look at the public health crisis and talk about lack of reciprocity in terms of people looking out for each other's safety, it's almost that when you invoke these things, in a context like the United States, people just don't hear what you're saying. The fact that somebody might not take precautions for the impacts of their own presence on somebody else's health, even if they know for a fact that they are probably not infected with the coronavirus, seems to show a missing layer of reciprocity there. So, just with climate change and with public health, there is widespread distrust, to the point where it doesn't really matter where somebody is speaking from. They could even be speaking from their own experience, having gone through the trauma of working in a hospital that is having a lot of patients in it, or speaking from a community that's having to resettle due to coastal erosion tied to climate change. Even people speaking from those experiences are not trusted by an enormous number of the population. And to get to that point signals to me that kinship is not taken seriously. It's sort of 'missing in action.'"

All of Kyle's points gather powerfully for Dahr, as a reporter of what was going on "behind the scenes" in the Iraq war, as well as through his extensive work in *The End of Ice*, which chronicles the radical environmental changes worldwide from a scientific point of view. He has been astounded by the enormity of a lack of care on both counts. He asks, "From your studies, where do you see the origins of that absence of kinship historically and geographically?" Kyle acknowledges the importance of the question, then proceeds.

"There are a lot of different ways to talk about the history, and the more ways that we can talk about it and the more diverse sources that we can appeal to, the better. One moment for me is particularly pivotal. I look at the fact that in the United States, for the bulk of the population, if you ask them to speak about the fur trade, they wouldn't have much to say about it.

"They might reference the Leonardo DiCaprio movie where he gets mauled by the bear. But even in that movie, I was trying to figure out where its setting was, and when it took place, and it banked on the fact that the audience isn't really thinking about the fur trade period and wouldn't have the same questions that maybe I do.

"People have a conception that there's a Native people before contact and Native people after contact, and especially for U.S. citizens, before contact usually means prior to the founding of the United States. But the point is that there was a centuries-long transatlantic fur trade era, prior to when the United States was founded. In the Great Lakes region, the larger group that I'm part of [Anishinaabe] was a very strong group at the time and exercised so much governance in the Great Lakes region that when the British got out of what's currently the United States, they thought the Great Lakes was going to be a large Native reserve. They didn't think it was capable of being settled or colonized.

"The fur trade era was a time of Native people responding to different forms of colonialism that were coming at them, reorganizing, and maintaining resilience. It was a time of tremendous violence, trafficking of people, and of slavery, and at the same time, there was also a kinship that was crucial to how Native people formed networks that made it possible for them to be resilient to all the changes that were going on. Native women, as well as White women who were part of Native families, exercised tremendous responsibility over trade networks, and had crucial leadership responsibilities.

"For them, it wasn't just about economic gain, but it was about actually maintaining kinship networks and exercising commitments to care and to reciprocity. And there's a huge literature that talks about the importance of reciprocity, the importance of trust within the fur trade period. During that period of time, which is roughly several centuries, you had radical reorganizations of society and tremendous expressions of Indigenous independence.

"But once the United States became the predominant settler power, and had a chance to figure out its agenda, you can see that it very carefully decided to completely get rid of kinship. The United States refused to negotiate with women leaders. American settlers expressed enormous prejudices against Indigenous forms of gender and sexuality. The idea of Native people maintaining sovereignty over areas like the Great Lakes was not something that they cared for, and they worked very quickly to dismantle Indigenous peoples' self-determination in those regions, which led to the forced relocation and resettlement of numerous Native people. The United States did not care about the fact that Native people had stewarded the landscape in a certain way to create an economy and a way of living sustainably that protected a holistic conception of well-being.

"For me, it's instructive to look at that period in the late 1700s and early 1800s, just to see how blatant the United States was—its private citizens, its government officials, its military—in dismantling kinship."

Listening to Kyle, I feel how important this information is in terms of taking responsibility for what is happening as a society today. Knowing specific motives over time sheds much light, and it takes where we are today out of an amorphous past and provides a possibility for change through understanding.

"I don't necessarily think that what I've shared is some kind of particularly early episode of it, but it's quite telling that you do have these moments where, very deliberately, something like reciprocity is, in a lot of ways, actually outlawed.

"This is important because when people read a lot of Native scholarship today, they'll see a lot of Native folks talking about reciprocity and trust and accountability, but oftentimes the histories we look at that talk about the growth of oppression and power, whether through colonialism, capitalism, industrialization, ableism, and patriarchy, don't oftentimes use the same terminology, and I think they should. We should be explicit that when we look at how capitalism works, at its economic exploitation, we can also talk about it as a lack of reciprocity. Is it spoken of as a violation of reciprocity? There are people that use that terminology, but I think we can use it a lot more commonly, so that these Indigenous values can be demonstrated as what are targeted by different systems of power."

I recall an article of Kyle's where he talks about the long-term nature of regenerating the sense of kinship in the face of those systems of power, and implementing it. Here is where our observations of the prevailing deep fears and reactions to them come in. In the talk I'd seen earlier, Kyle remarked that "the last person you want hitting the panic button in a bad situation is the one who caused the problem in the first place." That got a laugh out of my son, and because I don't want to see him living in constant fear, I remind Kyle of the confidence he had in Kyle's ideas. His approach was calming, an approach that opened everyday possibilities. In light of my son Rico's response then, I ask him what his thoughts are on the possibility to regenerate the kinship and reciprocity needed.

"That's a great question. And it's nice to hear that story. I hope Rico's doing well." He smiles, reflects a moment, and continues.

"One of the things I've been working on is the different ways of telling time. And this is all going to circle back to some of those earlier prompts that you've given me about panic and urgency. One of the things that gets underestimated is the importance of how we tell time. If you look at climate change or public health issues, they're almost always described in terms of clock time, linear time, a time where you've got equal units that

are sort of marching along. I'm trying to build off of decades and generations of folks that have been thinking about time.

"Here's the issue with clock time. Most of our day, unfortunately, is consumed by clock time, and that's not because any of us necessarily chose
that way of how we actually want to tell time. Climate change is always
about years and decades and centuries and how the changes are occurring. And it's also about the fact that time's running out. 'We only have
two decades,' or 'We only have fifty years' to do something. It's similar to
the public health crisis, where it's all about time. The fourteen-day waiting
period, or, we need to close the economy for a month.

"What happens is that this puts us in a mode where we are prone to certain emotional responses to very heavy issues, that perhaps make us pur-
sue solutions that are either not good or that aren't the best use of our
judgment, and that can be extremely harmful to other people.

"Just as a basic way of understanding how our conception of time impacts the actions that we might choose to take, consider a game of chess.
There is a difference between speed chess and chess you'd play where
everybody can take their time making their moves. If you're playing speed
chess, just the fact that you only have a certain amount of time to play narrows the solutions. It narrows the factors that you're taking into consideration. And it certainly doesn't make you feel like you have any responsibility
to anybody else, right?

"What if you were playing speed chess, and there were fifteen seconds
left and someone was trying to tell you that someone in your family was
sick, and you're thinking 'Oh, I have to make this move.' When you put a
ticking clock on anything you do, it completely changes your approach to
moral action, to action that promotes justice and equity."

Kyle's phrase, "it completely changes your approach to moral action,"
jumps out at me because we seldom see perceptions of time as influencing moral behavior. His thoughts show me how this fundamental
viewpoint of time influences our assumptions of what is moral, and
how it perpetuates many closely related faulty assumptions as well. He
goes on.

"What are the alternatives? Well, if you're not playing speed chess, you're playing where there's no clock time, so that completely opens up the type of factors you're taking into consideration. You're taking into consideration the fact that maybe you need to take a break because your opponent needs to have something to eat or you need to get a good night's sleep, or you can't play the chess game for too long or else it's going to affect your family life. When you're thinking about all of that, it opens up different possibilities that allow you time to reflect.

"You are still playing the game, but it's a different game at that point. There's a difference between a timed and a nontimed game. And we can think about climate change and public health in this way too.

"What are some of the alternatives to telling time? Can we actually talk about climate change through some other way? We absolutely can. There are a lot of different ways, and these ways are ones that Indigenous people have advocated for in different ways, using different language.

"For example, a lot of Indigenous people don't necessarily look at today's time as one century out of twenty or thirty centuries, in some precisely countable sense. We have a notion of deep time, the concept that today we're just operating in what must be a smidgen of all of the time that ever existed. And what is important is that when we talk about deep time, when we talk about the fact that we're currently living in a small period of time in relation to other eras, it gets us thinking about other things."

For me, Kyle is focusing on essential shifts of thinking needed in order to move beyond reactions based in a conventional sense of history. This is an expansion of what we consider our tools for problem solving in this context, an expansion into courting what we may not know unless we are willing to look honestly at potential limitations.

"When I say, 'Let's tell deep time through the last twenty or thirty centuries,' then that gets me thinking about recorded history. And it gets me thinking about the fact that most of the recorded history favors a European outlook and favors the outlook of powerful groups.

"If somebody were to ask, 'Did climate change occur during the Roman Empire or during the era of the Greek Polis?' I wouldn't know that because that's not taught to me in school. If somebody said the people responded to climate change in the second century, I wouldn't know because nobody taught me that.

"So I'd have to actually dig deeper. For Indigenous people, we think of our deep time, where we're thinking about the stories that we still keep today. We're thinking, 'Wait a minute, we need to go back. Is there guidance within those stories that can help us understand the challenges that we're facing today?'

"In that way, a conception of deep time makes you think that whatever we're experiencing today couldn't possibly be the first time it had ever happened. You should never presuppose that anything that's happening today is the first time. Maybe it is, but you should never presuppose that. You should always go back into that deep history—giving expression to the stories—which might involve talking to elders, talking to other knowledge keepers and knowledge gifters, going back into records of our storytelling traditions and really finding out: is this new in the twenty-first century, the first time that the Earth's temperature was rising on average, or the first time we need to worry about that? Deep time is one way of telling time, and one of the things that it gets us out of is the panic that whatever is happening right now is completely unprecedented."

Kyle hits home with this idea, and I think of all the people whose fear can be seen in the outcry "We've never been here before!" I hear it in the words and see it on their faces. Deep time gives another way to think about it all.

"Another way of telling time is kinship. Oftentimes I'll be where there is a Native person in a conflict, between them and somebody else, or between two other people. The person will resolve the conflict by opening up an entire history of each of the people's kinship relationships, both direct family and their societies. They could even go back centuries talking about people's kinship relations.

"They're seemingly not talking about time, they're talking about the actual relationships. Sometimes they can go back to clan relationships and nonhuman ancestors, and it's powerful when that happens, because those histories of kinship reveal that people are related in different ways, ways that the conflict that they might have today is betraying.

"This is not exclusively the realm of Native folks. What is important is that if kinship relationships are not just the responsibilities themselves, but are all the qualities that are attached to them, the reciprocity, the consent, the trust, and so on, then you can actually tell a whole history by looking at different areas of when there was an escalation of kinship or a de-escalation of kinship, when there was an expansion and the contraction of kinship. The fur trade wasn't something that we remember as three hundred years long; rather, we remember it as a time where there was an initial attack on kinship relationships through the first waves of colonization, then a revival of kinship. Then there was another de-escalation with the United States coming in, where again, you see a decline in reciprocity. You see a decline in trust, and then slowly you'll see different people beginning to intensify their kinship as a response to colonialism and further attacks on kinship.

"If you tell history in this way, over time it doesn't become surprising why it is that a wind power project might be considered okay to implement, even though the people that live on that land don't consent to it, or it might be possible to understand why people wouldn't wear a mask for the sake of being cautious and exercising some reciprocity, because what we've seen is a downward trend of kinship. It's not that there's a rise in global average temperature or rise in infection; it's that you see a decline in kinship and you see the markers of that decline, that people get infected when they maybe wouldn't have had to, or people experience coastal erosion when they wouldn't necessarily have had to.

"Looking at kinship also creates the sense in which the climate change crisis today is not the only time that certain people have imposed climate change on other people. The United States had a profound disrespect for kinship. They reduced a lot of Native people onto small reservations, didn't recognize as sovereign a number of Native people, and they had horrible sexist and patriarchal practices. They relocated Native people, and all of these are forms of environmental change, ways of invading somebody else's space and place that those people did not consent to.

"I think when Native people first started hearing about what a lot of people were saying about climate change and climate change being unprecedented in terms of a human society's capacity to inflict environmental damage or change, many people were saying 'Well, that's ridiculous. What about the relocation to Oklahoma?' This is because Native people are telling history also through kinship."

Kyle then introduces still another way to relate to time, a crucial measuring tool, and an inclusive view that implies a much deeper and broader sense of relationship and therefore, responsibility. It is a pragmatic view, going far beyond the viewpoint of one group of people.

"Another way of telling time is through the seasons. I've done some work with Samantha Chisholm Hatfield with elders in the Siletz community and other Pacific Northwest tribes. Samantha has spoken to a number of folks that talk about how Indigenous knowledge is about being able to keep track of seasonal change by being able to keep track of different relationships. During this season, you see this plant blossom, or you see this fish appear, or you see this animal go away. What's interesting is that people develop, over time and generations, incredible repertoires of all these hundreds of relationships.

"There's much that that means about one's relationship to the land, and some scientists will look at those observations and say, 'Well, they're not actually causal observations.' So scientists oftentimes don't know what to do with the knowledge because there's not necessarily a causality to the relationships, but what's important to note is that if you look at descriptions of how Native people actually talk about the relationships, the causality is actually not the point. When they're telling time through these indicators and relationships, they usually say things their way. They don't say that 'During this season, this blossoms.' They say, 'During the season when this blossoms, that's when people gather and do this,' or 'That's when people gather and do that,' or 'They might harvest this, and they might take care of something else.'

"They always talk about a social activity, and by social, I'm meaning human and nonhuman. They always talk about a social activity that has some importance. It could be cultural and it could be economic, so they talk about the relationship and the social mobilization that occurs along with it."

Kyle points again to the pragmatism behind this method of approaching time through expressing and following relationship. The language creates an altogether different sense of observation, and therefore another individual, emotional, intellectual, and community response to what is occurring.

"The idea is that all of these relationships are signals of when you do social activities that make up your entire self-determination, your entire society, and that help to protect the safety, the well-being, and the freedom of the members of the society. So, if you are following these indicators and you notice that one's off, you ask, 'Wait a minute. Why did this bloom a lot later, a lot earlier?' Your immediate thought is not, 'What's the biochemical origin of this?' Your thought is, 'Our social activities are off, and we need to all get together as a society.' You might need to meet with people in the same region who are in a different society to get to the bottom of this. Or, 'How are we going to reorganize our activities so we still serve the right purposes?' Or 'Why is it that the fish are not as plentiful, when they had been for a long, long time before, and we expected that? Let's get to the bottom of that. Maybe we shifted the timing of a ceremony, and then realized how that fits with all the other things that are shifting.' You get people together, and all of a sudden you're having a very different conversation.

"Notice that it creates a global response as a responsibility to address the issue in a way where people have to take everything into consideration. You wouldn't have what we have here, where somebody is alarmed by climate change, and it's said that we should end the coal industry, without seeing what happens to all those people that are employed in the area by that industry. They didn't choose that industry. They might say they have an abiding claim in it, but they didn't invent the coal industry. It wasn't something that they feel is a metaphysical truth. But, there's no caretaking for what happens to those folks that would be left by the wayside.

"It's like in those other examples of clean energy where people don't think about the consequences of their actions, because there is no tradition of telling time in that way. When people see a change that's unusual for them, they immediately want to start acting swiftly or ignoring it, or blaming other people. They're not necessarily interested in getting together and figuring out ways in which all of the interlocking pieces can work in harmony. To do that effectively, you have to have a lot of built-up reciprocity, trust, and consent. I think this is an important point about various tipping points."

To my thinking, Kyle is presenting the idea that the more profound "tipping point" is not 1.5 degrees Celsius, but the destruction of proper kinship relationships.

"So, if you take all these different nonlinear, nonclock ways of telling time together, you have deep time. You've got kinship time. You've got seasonal time. All of these other ways of talking about time can fill in a few gaps. What all these times do is get us focused on kinship, so the issue is not 'How do I reduce my carbon footprint?' The issue is, rather, 'Do we have the kinship needed to properly respond as a society, so that we won't mess each other up in the process?'

"If we talk about urgency and panic, those two words are good because when we panic about something, it's because whether we think about it or not, we're worried about two things. Something we've perceived not to be normal has happened and we didn't expect it to happen, but we also realize at the same time that we've taken for granted the conditions that made whatever was normal possible. When we panic, we realize that because we don't know the conditions that made our normal possible, we're going to do whatever quick fix we can to return to normalcy, even if that just means a sense of normalcy.

"This idea of normality, or the assumption of what's normal, is a key problem with clock time. No matter how we're dividing it up, it gives us a certain order of how events unfold, how long things take, and it gives us a sense that if some action is not done in a certain amount of time, the consequences are going to be particularly severe.

"That panic is focused on a restoration of normalcy without questioning whether what for us is normal is normal for everybody else, without questioning the actual factors that create that 'normal.'

"For example, with the public health crisis, we've seen that Native people, African Americans, and Latinx people have been affected severely by COVID-19. A lot of more privileged White people were also affected in different ways, but are adamant about returning to some sense of normal without understanding that their lesser risk of exposure or death or extreme sickness is caused by the factors that then make it so that these other groups are affected severely by it. And so again, people have shifted very quickly to talking about vaccines, to talking about their freedom to not wear a mask, to not take precautions.

"In Michigan, some of the people that were protesting to open the economy back up, had skits of people getting a haircut, where the effects on that industry and the restriction of not being able to get a haircut were assumed to be a paragon example of trauma and ridiculousness. In an example like that, people want to do whatever it takes to get back to normal, but it's not asking whether getting a haircut and cutting hair are what are at stake for everybody else.

"If we're 'urgent' about things, on the other hand, we're not focused on normalcy. In fact, I think when we are urgent to take action, we actually realize that whenever we fall into thinking that something is normal, we're probably wrong. For any group of people who is affected more by climate change or by COVID-19 than others, how could they possibly think that today's time is normal?

"Think about deep time. Would our ancestors—pick a generation—have thought this was normal? You can think about that in a lot of different ways. So why would we think that 'normal' is the right label to characterize a state of affairs that we're trying to get back to? People's use of that term, like 'When do we get back to normal?,' is very indicative of a way of thinking that is not going to be very responsible to the situations of others.

"Urgency is actually focused on kinship. For example, friends in the Detroit area talk about how it's not just the infections with COVID-19 that

are harmful to the African American community; it's racism. They're concerned about going to a hospital and getting infected with COVID-19, and
they may have another condition that would get worse or that may even
kill them if they don't go to the hospital. They're concerned about trust with
the medical profession. If you think about that situation, it's not about
whether it's normal or not; it's clearly not normal. What's important is, why
aren't people able to get the medical attention that they need?

"It is because the institutions have not behaved reciprocally, have not
behaved in trustworthy ways. And this makes it challenging to respond.
People are subject to higher rates of infection because they don't have any
choice but to continue working in the way that they did prior to the public
health crisis, because they are in different service and other industries.
Clearly a society organized in that way is not very reciprocal. There's not a
lot of consent happening in that society. And so in that way, urgency is focused on kinship and not focused on what is normal or not normal. Kinship is very skeptical of anybody claiming some state of affairs is 'normal.'

"As individuals, we need to take that perspective. Whether we're Native,
or someone from another group, or someone who's really privileged, our
today's times are dystopian times, they're apocalyptic and postapocalyptic
times, and that's not really a stretch to say that.

"For many of our communities, we were repeatedly devastated in a short
period of time, in the last couple of hundred years, and we are trying to
figure out what it means to rebuild the fabric of our societies. And not just
because we need to protect our cultures. What we're doing is protecting
our social nets, our networks of kinship relationships that will make our
lives better.

"When I advocate for tribal dignity, self-determination, it's because I
know that the families, or that my family, that my society, and some of the
related communities, are really my best bet if I want to have that network
of people that will give me the support that I need. And I know that my
support of them will be meaningful to them. That's why I advocate for Indigenous issues and the culture, the sovereignty, the self-determination,
the economic independence. If done through an intention to restore and

establish a kinship, they have benefits for us, and we are providing benefits back to others. If I have freedom, if I have safety, if I have well-being, that's not something I can achieve alone.

"Don't pretend that your freedom comes from your acts alone because it doesn't; it's related to all these different factors, and the better we acknowledge and observe those factors, the more we'll understand what our responsibilities are to others. With something like climate change or a public health crisis, no one individual causes it, yet the actual effects that it has will challenge our conception of freedom, challenge our wealth, our well-being, and our safety. That's not a time for us to then reinvest in myths, to protect some myth of independence, but rather to understand the first question, 'How did I get in this position in the first place?'

"You have to seriously reflect on all the relationships that got you to that point, and why it is that it's extremely challenging to respond, to mobilize people to respond.

"If we look at almost any dystopian novel, when we accept the fact that we're in a landscape where we have to recognize that due to some active other group, or some other thing, that all of our kinship network has been destroyed, then what does it mean to pick up the pieces in this landscape? What does it mean to begin to start building those kinship relationships?

"It's important to note that dystopian narratives don't end with the end of the dystopia. They end with the promise that somebody has developed greater kinship. The group is greater at a small scale. And that's where there might be something that we can think positively about, even though the dystopia will persist. What that means is that we need to completely transform how we value what actions and behaviors are conducive to responsiveness to climate change and public health.

"There are people who are doing that in the context of their household. There are people that are doing that in the context of their job. There are people that are doing that at the international level, but which ones do we see on the news? We only see a few of those. Except in rare cases, we don't respect or give awards to people that work at the household or the neighborhood level. We don't actually tend to value the diversity of kinship work that people are doing, and hence it becomes very tiring work and difficult work because it's not actually considered to be a pathway for somebody to be creating change.

"But it's a huge asset. It is. And I think to understand that, you have to also be in a certain kinship relation with others. If I ignore the contributions of others, it's probably because I don't know how to listen to others. In this way, we're seeing the ripple effects of how multiple layers of kinship are putting us in a space where it's even difficult to acknowledge that there are diverse contributions out there to resolving these issues and to really think cohesively about how it all might come together.

"It challenging because it's 'starting small,' in American English. Starting with your daily relationships is extremely important, but that language already minimizes what it would actually mean, to have a view of a kinship system where you understood exactly how the social activities of different people in groups and communities fit into a responsiveness. That is one thing that's challenging and difficult to articulate."

For Dahr and me, Kyle's collective points, and the trajectory of his discussion, precisely address why it's important to gain the perspectives of many different people. I tell Kyle that "in Dahr's presentations on *The End of Ice*, many were saying, 'This is overwhelming. And it's the end of life as we know it.' So in response, we're gaining insights from people of all different parts of society, all different age groups and expertise, finding out what they're thinking and doing. We're hoping that this will generate what you're talking about, which is a lot of voices talking. You're saying it's thorough communication that needs to happen, and one of our younger contributors, Raquel Ramirez, says, 'Listening creates compassion, and therefore better decisions, both individually and collectively.' She says that 'communication creates understanding.'" Kyle listens closely, and offers his observations about information and how it is put into the world, on how it functions.

"In journalism, they'll talk about particular examples, of how some tribe put up sixty solar panels, and the way it's written suggests the tribe has solved the issue right there. Those sixty panels evidently power the whole tribal enterprise and its four thousand members."

Kyle's point makes me think that while the article wants to sound a positive note showing a solution, there may be many assumptions made from it as a partial communication not fully addressing the balance between the project's real capacity and the need. The article may also be a reflection of the writer's fondness for solar power. That predilection and those assumptions can have consequences running counter to what the article seems to want to convey, so Kyle explains how this might work, and offers alternatives that reflect Raquel's point about really "listening."

"That is actually presenting profoundly problematic ripple effects. We see somebody put on a pedestal and think, 'Well, I can't do that where I am.' However, I've seen journalists and academics do things a different way, where they're able to contextualize how a particular case or action fits into a larger fabric, and this helps people to understand how they can feel better about what they might be able to do."

Kyle's points remind me again of concepts of time, embracing inclusive priorities, and I tell him of my experience with a Native group where solidarity between autonomous individuals was crucial, where it was the absolute priority, and a meeting had to include everyone's thoughts, no matter the amount of time it took to accomplish that. He immediately expands on the thought, providing a kind of ultimate context that forms our daily lives, perhaps without our reflections on it.

"Exactly. And one of the worst forms of clock time that biases what we think, and this kind of conception obviously depends on what community you're from, is how long each of us is going to live.

"I think I'm going to live a certain period of time, and I think of what I am going to accomplish in my life. If I'm healthy, I'll probably be somewhere between fifty and a hundred. But I think that if I don't get stuff done by sixty-five or seventy, I'm not going to see any change. Where did we get that assumption from? That's how long our lives are going to be, and that creates a problem.

"It creates a sense of panic. It creates a sense that even for people that have had a lot of positivity associated with their work, they don't feel they're given respect for what they've accomplished. And we don't have abiding traditions that really understand that we're planning for the next three, four, seven generations, for the long term, and what it means to think about our kinship as transcending whatever we perceive to be our life.

"I want my kinship to be whatever kinship I provided as what ripples out. In being recognized for work and getting respect for our function, and so of kinship, we need to think about what type of team we want to be part of. We want to figure out who the collective of people are that are meaningful to us, that are responsive to change, and think about what it means to build that kinship network. What's interesting about kinship is that it's slow, because we have to be patient with the development of the relationships.

"But there are also times in which there can be great leaps. If I have a strong kinship group with diverse people that I've worked with for a long time, that might allow us to work with another group that maybe is from a different area, and because all of us are strong in our kinship, we trust our leaders. We know that they value our consent, and so they can actually link up with others in the network."

Kyle's point about choosing the right team is critical, so he focuses further on that, showing what is possible in a good way, which can be seen through examining Native relations over many generations prior to colonization, and as a response to it. However, he also wants to point out that we cannot overlook all the possibilities of what kinship might mean.

"Kinship, especially given how it's been damaged and erased so much today, also makes us actually forget that our ancestors, without any of the technologies we have for communication, and continental and even hemispheric networking, solely through kinship relationships, without the various technologies for looking at regional and continental, were actually able to generate timing, trade, cultural exchange, linguistic exchange, solely through kinship. That's the power of kinship, but kinship is one layer of our lives, and it can also be an area that's very harmful to us.

"Often people we consider our kin are the ones who are in the position to be the most abusive. Often it's our kin who we feel motivated by to be abusive to ourselves. All these other aspects of our relationships with others, including rights, contracts, and other aspects of our relationality, do have a place. But we're living in a situation where there's been such an emphasis placed on relationships that have to do with individual ownership of property, individual rights, that other aspects of our morality have gone away. We're seeing a world that reflects that, that dominates with a certain narrow view of kinship. A nepotistic society is a type of kinship, but is completely unbalanced.

"Here there's a line to walk, where people have to recognize that we're not arguing for kinship because we have a bloated view of what it is. We're being very careful in our interpretation that a lack of kinship and the defamation and destruction of kinship is a particular problem for this state of affairs."

At this point, Dahr offers that for him, Kyle is talking about standards of living and perceiving, about the nature of relationship. He remarks that Kyle has woven together the questions we've asked him clearly and intricately, and he broaches a vital remaining question for himself. Thinking of a previous contributor, he asks Kyle if he can address his concern: "How do we comport ourselves? One of the people we interviewed, Ilarion Merculieff, talked about the importance of that, and of not being attached to results, because colonialist culture is all about results. Again, it's like reverting to how things actually should work, which is about whether we're looking at time and kinship in the right ways, in the right perspectives. What you've been talking about is the right conception in relationship with time and then kinship and how everything comes from that. So, how do we move through this in the best way possible?"

"Whether it's going in a better way, or going further into what's gotten us to this crisis, I don't know, and we're all learning," Kyle replies. "When we take actions, I think about the learning that I have had to do.

"For example, in relationships with students, students have distrust for faculty in some programs, especially where you have higher education programs where there are issues of risk, like of sexual violence and harassment of different kinds. Students might say, 'In this department, our consent or voice doesn't matter,' or 'We don't have any trust.' And the faculty will say, 'Okay, let's create a policy that protects your consent, or let's create a policy that vouches for the trustworthiness.'

"We think that we can institute interventions and have people follow them. But oftentimes we'll get into situations where we've come up with the best policy for students to trust faculty, and the students will say, 'Well, that's not enough.' Or, 'We still don't trust you.' And so oftentimes people resort to anger or frustration or giving up when they hear that, but they haven't grasped that the issue is that generations of abuse don't resolve in 'Here's the solution.'

"It's similar with work in climate change, when somebody wants to put in a hydropower dam and they consult with people that live there, with Indigenous people, and those groups say, 'We don't trust you. We don't, even though you've had some meetings.' Look at the Dakota Access Pipeline. The Army Corps of Engineers had consultation meetings, and the company did too, but the tribe and other folks said, 'That's not consent. That's not respecting our self-determination.'

"Again, when we're not focused on kinship relationships, those things appear as frustrations to us, as things to be angry about. And it can become very hard for people to think about their actions.

"But trust or consent aren't roadblocks. Those aren't hurdles. Those are the point. And what if that means that there are certain things that we might not be able to see stopping because we focus our attention on establishing restoration of kinship? I think we might have to be okay with that."

This thought brings back the emphasis that we are in a long-term game, that we are not in a timed chess match, yet we are dealing with many players who are seeing it very much from a timed perspective, and with the clock running. In keeping with his admonition to keep a sharp eye in such a context, Kyle returns to emphasizing that we need to continually and honestly observe the nature of the team we choose in our pursuit of proper kinship.

"What if that means that the U.S. government invested its time in the cultivation of kinship? It's very scary to think about this. That might mean the progression of harmful activities, but how are we going to process that? How are we going to think about that in a way where we're thinking about all the different voices that are involved, and about all the different ways in which people are impacted?

"It's a huge challenge and has a bit of circularity to it, because since we don't already have sufficient kinship with others, that means we oftentimes don't get the communications that we need about how others are affected by our actions. And so when people are cautious, when they're slow to act, when they express that trust represents a huge issue and hang-up for them, we need to take a step back, not look at that as a hurdle, or even as a problem in the sense of one that can be solved. We need to look at the kinship network issues and think very carefully about what the actual significance of our behavior is and what our behavior actually means for the long term."

These thoughts conclude our interview, and I tell Kyle that what my son Rico felt that day with him is borne out in the balanced approach of this talk. Rico felt an attitude, and he trusted Kyle's approach to seeing the many sides of our situation. The trust is what it's all about, and we can't create a policy around that. We can't institutionalize that. It's something we have to live, and we have to want to live it. I convey this to Kyle, with gratitude.

"Right. That's exactly right. And thanks again for sharing that story of Rico. That's awesome that we made that connection there."

Dahr says, "We appreciate your time, Kyle. It's a delight and an honor to spend this time with you."

"No, it's an honor for me," he replies, with a nod and a smile.

7

Terri Delahanty (*Cree*)
Sacred Feminine and Sacred Masculine

COMPOSED BY DAHR JAMAIL

We need to get back to where there's no separation between you and me. I pray for it, because I think that's the only thing that's going to heal the Earth. I pray for all those unborn generations, for all those new species that can come about.

—Terri Delahanty

Terri Delahanty is a Cree woman who maintains a regular practice of, in her words, "Native ceremonies, meditation, and women's traditional ceremonies." A Sun Dancer of eleven years, Terri has also been pouring sweat lodge for two decades, and is a pipe carrier.

"My spirit's journey is bringing knowledge to the community about returning to the Sacred Feminine and Sacred Masculine that has been lost through patriarchal mind," she told me, when I asked her to share a little more of her background.

A founding member of the board of Women In The Spirit, Terri also sits on the board of trustees at the Institute of American Indian Studies in Washington Depot, Connecticut. She is also on the planning committee for the Wisdom Keepers Gathering in Kansas, as well as being on the panel of elder speakers and emceeing the gathering.

An ordained minister, Terri is the Native American chaplain for York Correctional Institution, a high-security women's facility, as well as for three of the men's prisons in Connecticut, where she offers sweat lodge, group, and one-on-one counseling, in addition to ongoing traditional ceremonial teachings. While federal and state institutions use terminology such as Native American, Terri pointed out that she prefers referring to herself as Native or Indigenous. She also facilitates groups in the Red Road to Wellbriety, a 12-step, evidence-based recovery program.

Following her Native traditions, as she put it, "brings richness of Spirit into my professional life." Terri has been working in an educational setting since 1991. She is a certified parent educator, supervisor level, through the National Parents As Teachers Organization; director for Greater Hartford Even Start, a family literacy program; and director for the extended day program and program coordinator at the University of Hartford Magnet School for twenty years. In addition to all of that, Terri also worked as a braille instructor for nine years.

In October 2019, I spoke with Terri in her home, the day after sweating with her community in the lodge she built in her backyard. She pours sweat lodge twice a month, while the other weeks someone is pouring lodge for her community elsewhere. One of the lodges she pours is for the broader community, while the other lodge is just for women.

While we spoke about the importance of non-Indigenous people having a deep understanding of what Indigenous people have faced throughout their colonial history, Terri spoke about generational healing. The energy from the sweat the evening before was still very much with us, and we wasted no time going right down into the heart of issues she wanted to discuss. Terri looked overseas as well, to other victims of genocides, like Jewish people, in addition to Native Americans, to acknowledge how people who haven't experienced such atrocities might not have the capacity to share the deep understanding of what has really happened to Indigenous people or any other culture that has experienced genocide.

One of the core themes of focus for Terri was the sacred masculine and sacred feminine, because, as she put it, "We've gotten away from our heart center." As she sees it, this is because, without the ability to "sit in our hearts," there really isn't the capacity to understand, let alone empathize, with a people who have survived genocide, or the slaughter of six to nine million women across Europe who were accused of being witches.

"Women were burned, you know, and I think of that sacred masculine that couldn't protect their women, or the families that were separated because in order to protect this one, you had to give up that one," she explained. "And so where's the family unit?"

She sees the internal spiritual rupture and wounding of the sacred masculine and feminine that occurred amidst the slaughter of so many millions of women during that time as an open wound that still needs to be healed.

Terri also believes humanity has reached a place where people need to attempt to understand the Indigenous cause because, as she put it, "'Indigenous' is a place."

"We knew how to have that life source, that sustainability for over a thousand years, ten thousand years, twenty thousand years, or more. So I think if people come to that understanding, then maybe we can live," she paused, the emotion palpable between us at the kitchen table where we spoke, before she said her next words, slowly and with her voice lowered, "as a community."

She paused and looked out the window for several seconds, then continued.

"We're born of beauty for a thriving life, and there's the sacred hoop. We all have a place on that sacred hoop, and there are elders that know where the giraffe is, where the water is, where the tree is, where the humans are. If something marks the surface, and if one part of that doesn't live up to their part, the hoop starts failing. And that's what we're in right now. I really think it started with humanity in every culture. There was something broken in the heart, and until we come back to being heart-centered, how do we understand each other?"

Terri spoke to how, as she sees it, even if non-Indigenous people are unable to grasp what Indigenous people have gone through, "There's something about the trying that seems really important, nonetheless."

This is because, as she sees it, if humans are to have a chance, we must all return to Indigenous ways of being. "Because Indigenous people understand their place, they know how to live in that area. There are Indigenous here that know how to live in this area with the waters and the mountains or whatever the landscape was right for, sustainably. So I think we're coming into a point where Earth is doing what Earth is doing because we're not together as humanity."

For Terri, Indigenous people are all about community, but when she looks out at the world today she sees a glaring lack of community. She sees fewer children playing outside in the neighborhoods, fewer gatherings (we spoke several months before COVID-19 began ravaging the country), less support within communities for one another. She understands deeply how none of us can get through life alone.

"I didn't receive what I know from Grandfather Lloyd sitting here by myself," she said. "I was in front of the community. I didn't receive those eagle feathers one-on-one, it was in front of the community because community is the one that helps me walk with the eagle feathers in a good way and stay on this path."

For Terri, this is her way of having her Creator see her walk in beauty, when she is living her ways within her community. "If I'm walking in beauty, then you can darn well be sure that my community has helped me do that, because I can't do it alone."

Speaking pragmatically about what needs to be done, Terri came back to the imbalance between the sacred masculine and sacred feminine.

"The sacred masculine carries that fire, and the sacred feminine carries that water," she explained. "And for centuries now that masculinity has been drying up the women. If you talk to many Indigenous cultures, it is about really coming back to that balance of me not putting out your fire, but also you not drying up my water. It's so out of balance now, I think it's necessary for communities to come back to honoring the women and the fact that we are the bearers of life."

Terri spoke to how centuries ago, when a woman was on her moon time, she would literally sit on the Earth, and the sacred masculine was her protector and provider, so she was allowed that time to sit on the Earth. Only by way of that tradition could women develop different ways of knowing, through sitting on the Earth month after month, year after year, decade after decade.

"That's been stifled," she said. "We need to come back to really figuring out what the woman has to offer. And I don't think humanity is going to solve anything until the woman steps back into her power."

Terri pointed out that she in no way means "power over," but simply that women are recognized for who they are.

"Elders say the same thing," she added. "I've never been in an Indigenous ceremony where the woman wasn't honored for being the life bearer, the life giver, and all those ways of knowing are because the male is the protector of the universe, right?" [She then laughed.] "It's the woman who is directly connected to the Earth during her ceremony time. The Sun Dance is about praying up to the universe, to the Star Nation and Creator, it's all about upward and outward. But when a woman is on her moon, it's literally inward and downward. She's literally flowing downward and receiving that nurturing and that nourishment from the Earth."

Terri went on to speak of how we each have both the sacred feminine and sacred masculine within ourselves. In this way it was possible, for example, for Indigenous women to know which men were going to become chiefs, and which were to be warriors, similar to how men knew which herbs to pick, and the women would go pick them.

Similarly, her concern about the power-over dynamic is that it isn't just men doing it, but some women as well, because it's not just the masculine causing it, it's more the thinking that causes this problematic dynamic.

"So how do we get rid of that power-over attitude," she asked. "It's about really honoring what you bring as a male, and what I bring as a female. If we can bring that into every relationship, then our children are going to see that we're going to be honoring our elders."

Terri shared all of this, then came to a specific point.

"It's about coming into ceremony," she said softly. "It's about coming to ceremony and partaking in how we pray, what we pray for in that connection to the Earth. We're not connected to the Earth. You see what's happening because we haven't been connected to the Earth."

I asked her to speak more about ceremony. Terri struggled to find the words.

"Language just doesn't suffice for what it brings into my being," she said. "It's not about peace, calm, or connectedness. It's all of that, but I think ceremony for me really puts me in alignment with Creation and Source and there are no boundaries to my physical form when I'm in ceremony."

She went on to say that whether she is in sweat lodge, or conducting hemblecha (vision quest), for the last decade she has been asking Earth what She wants her to know.

"When I moved here, I didn't just put up a sweat lodge because I wanted a sweat lodge," she explained. "I didn't decide to pour water either. I was asked to pour water. The Grandmas put me there. And I said no [to the sweat lodge, at first], because I know what it takes to carry being a water pourer."

But she's been pouring water in sweat lodge ever since, because she was told to do so.

When she moved to her current home, she noticed how every time it rained, her backyard would flood. She kept offering tobacco toward her desire for a lodge. One spring her backyard flooded again, so she called her elder from Wisconsin, Bruce.

"Bruce came out and offered tobacco and we sat in prayer," she said. "A week later another layer of ice melted, but the water just separated, leaving an area where there was enough room for a sweat lodge and a fire pit near it."

She looked at me and smiled, then continued.

"So that water responds to the prayer. Bruce said, 'Now Spirit's here. It's time to put the lodge up.' So that's when it started."

The same thing happened to her at her previous home. She didn't know if she was going to have a lodge there. She prayed and offered tobacco, then went up into the woods with her partner at the time, where they saw a deer. They stopped, held hands, prayed, and offered tobacco, offering to serve the people there. Bruce came out and offered prayer at her house while Terri was in New Hampshire making a drum with her nephew.

"I got a phone call from Bruce," she said. "'You got to get home, Spirit's here. It's time to build the lodge.' he tells me."

It's only with permission that she builds a lodge, with Spirit being present. Otherwise, as Terri put it, "it's a dead lodge."

The message she has always gotten from Earth is around water. Terri grew up on the banks of the Farmington River, which had the great flood in 1955, and the following year she was born. "So I believe I was born of the water," she said. "When I was a young child, I was always down at that river and I brought my whole life to that river."

She speaks of the understanding that modern science has finally come around to, which is that water holds memory and emotion. "Sure, we know a lot about water, scientifically, but we'll never know everything there is to know about water," she said. "We pray out loud so that the Grandmothers and the Great Grandmothers and the Great Great Grandmothers recognize the tone of our voice because that's where we received the tone in our voice."

She smiled and looked at me for a moment. "Your Great Great Grandpa recognizes the tone of your voice, Dahr. So in our tradition, shouldn't we pray out loud? I would always go to that river and offer my prayer or offer a prayer song. Then years later when I'm up in the woods and I'm doing my hemblecha, I believe that that tree knows me a little bit more. Because water is constantly traveling, whether it dissipates and goes up into the clouds, and then goes over Alaska and drops as snow or rain. It's always moving, all the water that's here is here. I believe when I go out into the woods and I'm not on that river, that bear that looks me in the eye knows me a little bit better."

Terri's practice in serving means she consistently asks the Earth what it wants her to know, what it needs her to do.

"It shows me it needed me to put up a sweat lodge to service the people here," she said. "I know that I have been attuned to water my entire life. It has always been the water, specifically my connection with the Farmington River, when I've had great change in my life; whether it be success or grief, sorrow, or joy, it has always been that river that has guided me."

Terri believes when people come to her and others' ceremonies, they will learn more about their connection to Earth, and from that, how to honor each other. But when she looks out at the world today, she is dismayed by the level of intolerance of one another.

"We need to understand and honor each other, because what we don't understand, we fear, and what we fear, we destroy," she said calmly. "Look at what was done to the Jews, to Native people, Black slavery."

For her, that understanding comes from listening.

"If you can just deeply listen, even if for just twenty seconds, all of a sudden you start to understand the other person's culture, and you are partaking in it and not putting into it what you think you already know," she said. "You learn something new, and then you understand. Then you don't have to tolerate, because you *know*."

I asked her to share her thoughts about the climate crisis.

Having been to Standing Rock during the attempt to block the pipeline, Terri met a man from Canada who used prayer and ceremony to stop a pipeline there.

"Prayer and ceremony can make or break anything," she said of this. "And I'm at a point in my life where I don't even need to believe any more, because I know it's true. Ceremony has changed my life, it's changed my behavior, it's changed the way I treat people on any given day."

She believes if more people aren't connected to the Earth, there is not a way out of this darkness. "If there is no connection to Earth, and we're not bringing our children there, and we're not going to see that the words from our elders are right, I don't see a way out of it."

I asked her, then, how do we live during this time, and does she see a chance for the reconnection of which she speaks?

"Even though Standing Rock failed to stop the pipeline, I drew hope from it because of how Indigenous culture, prayer, and ceremony became mainstream," she explained. "Indigenous culture was seen and acknowledged there, and the non-Indigenous people, as well as the media, saw it there, and saw how opening your heart and bringing prayer into your life works."

In a sense, it was an example of the listening to which she spoke. For a brief moment, Indigenous people were listened to at Standing Rock. If people's eyes and ears could be opened at Standing Rock, then "their hearts will be touched," she said. "I just want to be in ceremony more because I know the power of prayer. I know the power of ceremony and I know how many people it can touch. It can change behavior."

She knows this because she's seen it in others, and has experienced it herself.

Reluctant to talk about herself, however, she cited her friend Scott, who was in the lodge with us the previous evening, and who is now walking the Indigenous spiritual path. In fact, everyone who attends her lodge is non-Indigenous.

"I am seeing a profound opening of the heart. Sitting in the dark in the lodge, the drum is going, the songs are being sung, and people feel the connection to the Earth. People are asking ancestors for guidance," she explained.

Terri said she believes, whether people admit it or not, that we all know we're headed toward extinction. "This turmoil of extinction of things is erupting, but it's also somehow connecting people to the Earth," she said, holding her hands up in the air. "I don't even know that they know it's connecting them to it, but more of them are coming to the Indigenous ways to find peace in their life."

Knowing our time was growing short, after sitting for a moment, Terri found it necessary to speak more of the sacred feminine and sacred masculine. She repeated how she does not see how humanity will survive this crisis if we fail to return to the sacred masculine and feminine.

"Women are destroying themselves in the dominant culture, and are destroying other women, and until we come back to the sacredness of being a life bearer and a life giver and all those ways of knowing that accompany that, we are lost."

Terri believes if women are not supporting each other, they are not going to know the sacred feminine.

In the sacred realms of which she speaks, she explained that men are physically stronger and hold that power in a good way, as they are the protectors of women, and can be providers, not just in a financial way. One way of providing of which she speaks is adoration.

"When you adore the woman in your life, she is filled," she explained while smiling. "And when we adore the man in our life, we are filled, and then we become a unit. But we've gotten to a place in the dominant culture where we just try to outdo each other, and this is reinforced by society and the media."

She talked about how media projects the image of the perfect female body as the norm, meanwhile most women remain clueless about how to honor their moon time.

"They're totally clueless," she said, laughing. "They walk around, they have to hide their sanitary things for their moon time. But there was a time when the sacred masculine would allow us to be in our own ceremony. The women that were mooning at the same time would be in the teepee together, dreaming dreams. And a lot of times they would dream the same dream and come out and give it to the community."

Then the men would take actions based on these dreams, because, as Terri said, "Then the sacred masculine is taking the right action, and the feminine is the emotion of it. And you put that together and there's the sacred masculine and sacred feminine living together in the proper way that we were designed for in order to live a thriving life."

She sees us as having stepped away from this, in the way that started long ago, which she had spoken of.

Terri lives the sacred feminine. Ever since she's had her own children, when her sons and grandsons come into her home and tell her they are hungry, she never tells them to wait until dinner. "I also taught them that prayer is in this food," she said, smiling and pointing to her cupboard. "There's prayer, and it's going to nourish you and it's going to nurture you. And I think that's part of that sacred feminine, bringing that prayer to the young boys."

Terri was saddened by how many women have become uncomfortable with that role.

"I love taking care of my home," she continued. "I loved my home when I was married, and I love it when I'm not married. I think that taking care of the home is part of being in the sacred feminine. I don't know why people shun that, and I think it's degrading to do that.

"We need to go back to really learning what sacred feminine is, what sacred masculine is, and allow our men to be men and not try to be the strength that *they* are. We can let them take care of us. The sacred masculine can protect and take care of the sacred feminine while she takes care of the home.

"And that really is literally the perfect analogy for the entire planet, right?" she asked. "And instead, now, essentially, almost nobody's taking care of the home. Everybody's just out running around making money, chasing things, addicted, distracted, and completely missing the point."

While they were "dirt poor" growing up, Terri remembered watching her mother, who always had something to cook, and how much her mother loved taking care of her six children and husband.

Her mother taught her about compassion, along with how to read, to look people in the eye, and that what you have or don't have is not important. She also remembered her mother always watching the news, since her brother was in Vietnam. Terri would watch her mother weep when the news showed the deaths in Vietnam. "My mother would say a prayer for the person doing the killing, not the one who was dead," she said. "That is a woman with great compassion."

Terri came into talking about the importance of honoring her Indigenous elders.

Her eyes welled up with tears while talking about watching the elders in her life. "Every time they offer a prayer, they weep because they know how far out that prayer travels and how it is a prayer for all beings."

Terri spoke of them reverentially, her speech slowed, and she was speaking even more softly. She knows that every time she prays, she can tap into her elder Grandfather Lloyd. "He wept every time he prayed because he knew the power of the prayer. And Grandmother Bertha Grove, she was so ancient looking, with these craters in her face, and just being in the presence of an elder like that creates no separation. That is a true elder."

Looking back now, writing this in 2021, it was a foreshadowing of wisdom to be used for what was to come. In a dominant culture that is now more separated than ever, and with this exacerbated by the global pandemic, Terri's words about the importance of not having separation carry all the more weight.

"We need to get back to where there's no separation between you and me," she said, very softly, yet earnestly. "I pray for it, because I think that's the only thing that's going to heal the Earth. I pray for all those unborn generations, for all those new species that can come about."

She believes if we all do this, it is still possible to come together, but not without the elders. "We need more of those elders who are dying so rapidly," she concluded. "We need to sit in their presence, and we need to gather as many people as we can to sit in the presence of the elders, because they do not create separation."

This is the fundamental learning Terri sees us in need of as a species: knowing we are connected to all things, to one another, and to the Earth. When that happens, we will then behave accordingly.

8

Steven Pratt (*Amah Mutsun*)

Returning to Fire

COMPOSED BY STAN RUSHWORTH

As a collective, we can make this go in a better direction. And everyone has their part.

—Steven Pratt

Steven Pratt is a twenty-four-year-old student of environmental science at Cabrillo Community College in Aptos, California, and a member of the Amah Mutsun Tribal Band, his late father's people. The Amah Mutsun count their origins as being of this very land. Like many of the local Native people in this area, Steven was not born here but in the Central Valley. Families who survived the local mission found agricultural work inland after the mission system was abandoned in the 1820s, and after they were further displaced from the area by the Anglo settlers in the mid- to late 1800s.

"I was born in Sacramento, where I lived for about eight years, then moved to northern Idaho for another eight years, then here to the Santa Cruz/Monterey area for the past eight years, increments of eight." He chuckles at this cycle, smiling.

"My Native side is my father's, native to the Monterey Bay. But he wasn't really involved with the tribe much." He describes an early separation from his father, then connecting with him later in life, but losing him three years ago, when he passed on, early for his years. "I wish I'd connected more with him. As I started reflecting on who my dad was, with my Native background, I had the feeling to get involved with this side of the family, and to find that connection that I was missing from my father."

Steven describes feeling lost, looking for direction. "I'd dropped out of college. At that point, I was just seeking purpose, especially after my dad passed away. I was looking to find meaning in my life." He speaks quietly, earnestly, pausing to reflect as he forms his thoughts. Then he comes back to what he says was a critical choice.

"I chose to go down this path because it felt like the only option. It felt like I could either go inward, or go outward and seek out my Native family. And so that's the choice I made. I found my way to doing a prayer run for Indian Canyon [a small, local nonreservation Native community with four-thousand-year-old roots on their piece of land], a nineteen mile run, and that started everything.

"I ended up meeting a lot of really great people. And I found my way to sweat lodge through that experience because one of the runners was a sweat leader. He took me in and I sweat with him for about a year, and it helped integrate me into ceremony and culture.

"Sweat lodge was really the first experience where I got to embrace Native culture, and with a really good community. Through that, I was able to find Val Lopez [chairman of the Amah Mutsun Tribal Band that Steven is affiliated with]. And that's what I was seeking. I wasn't seeking money. I wasn't seeking anything along those lines. I was seeking that family and interconnectedness with our ancestry and heritage and tradition and culture. That's what I was really looking for, so I got invited to a ceremony on Mount Umunhum, and through that I met Val.

"Later on, Val invited me to a wellness meeting, and I learned about the stewardship program our tribe has, the Land Trust, revitalizing culture and Native practices that our family has done for countless generations. And though I was already working, I decided to leave my job to do that work."

Steven pauses here, then a broad smile crosses his face.

That work "helped bring me to who I am now." He stops, then reemphasizes where he had been in himself. "Before I joined the Land Trust I was really struggling to find my purpose in life, what I wanted to do. And when I told Val that, he said, 'You'll find your purpose in the Land Trust.' I didn't really fully understand that at the time, but I rolled with it and I signed up. What that led me to realize is how important it is for me to steward the Earth, and to really work with the Earth."

Steven leans forward with a look of satisfaction. "I found a really good groove in it." He describes one of the main projects he's working on now, "a 'killer' State Park project where we're revitalizing a coastal prairie that's become encroached with Douglas fir trees and has turned into a thick forest." He describes the hard physical labor, the chain saws, learning all the tools and about the burning needed, then pauses to add, "But what it also opened up is learning the language. I don't know much, but I'm learning, and it's really incredible being able to hear these words that our ancestors have been speaking for thousands of years, to be able to learn how to actually say them. It holds a really deep meaning within me, though it will take time to fully learn the language. But I know that it's on my path, and I will learn it."

He circles back and reminds us again, "So I came from a place where I was seeking, trying to find purpose, and when I found my way to this Land Trust, I found something that I cared deeply about. I was able to find a purpose because I really care, and to take care of the Earth is a blessing and an honor!"

Dahr says, "Beautiful. Thank you very much. You've touched on a couple of things we want to come back to, like the value of ceremony in your life, but before that, do you have anything you want to say about climate change? Do you have any comments you want to make before we get into specific questions?"

"Yes." He takes a breath then speaks with authority. "The thing with climate change is that the climate has changed historically. The world is constantly changing. But the difference between this climate change and the other ones, is that this one is the result of human behavior.

"That's the difference. And that's the thing that needs to really be highlighted, and although it has been, I don't think people fully realize it. Those other changes were natural, a result of Earth's processes. That's just how the Earth works, constantly changing, but we've found a position where we're manipulating the Earth and its processes, and that's causing the whole world to shift."

I ask Steven what he sees as causes.

"There are two things I notice. One is greed and the other is the lack of community. Those are two big things that I see a lot.

"I say greed because the robbing of resources has been huge. One thing we can all see is the overuse of oil. This has robbed from the Earth, and it's disrupting places, especially impoverished areas where there's a blend of oil and big companies that come in, damaging the people and the land. And it happens all over the world."

Steven pauses, then continues.

"The reason why I say a lack of community is because of how much individual isolation is happening. Historically, we've been true tribal people, or at least we've lived in communities, groups, and we worked together to survive."

He details large and small ways in which people today can stay removed from each other, cut off from the sources of their food. He says people do not know the origins of what they eat, and just put food in the refrigerator for days without having to go to the field or market. "We don't have to depend on each other."

I ask him how he feels we should respond to those challenges.

"That's really difficult, but at the same time, it's simple. You have to surrender, in order to see what needs to shift and how to shift it. Ultimately the biggest way to respond is to be a steward of the land and the water, doing everything to take care of Earth. And we can individually do that. Every single person can make a huge difference just by taking care of the land and water they occupy, whatever water source they use, and whatever land they live on. They have a responsibility. It's not someone else's job. It's all of our job collectively. But you have to allow yourself to become aware of your surroundings, and that's what I mean by surrendering.

"The difficult thing is that people tend to not want to face that truth, and this brings it back to the isolation, to when you're in your own little bubble. It's the classic scene of where ignorance is bliss, but being isolated is creating a false sense of happiness. So again, in order to get to that point, to where you can steward from a good place in your heart, you have to be able to allow yourself to let go of that isolation and feel the interconnected community that we are, as a whole human race."

Listening, I reflect that what Steven is describing takes courage, and it leads me to ask Steven what we're asking everyone: "How do we comport ourselves as we move forward? What are the factors that keep us from doing exactly what you're saying? And how do we respond to those impediments?"

"That's tough," he says, shaking his head. "Because it's about starting to realize that what we do right here in our yard is connected with every-body else. Everybody, and this means including *ourselves.*" He speaks the last word emphatically, as though bringing us into community with the statement.

"And, how do we carry ourselves? That itself is another difficult thing because in order to carry ourselves forward, we have to change, and that's hard. It's hard to change because we build a kind of exoskeleton that keeps us in our place." He speaks with personal authority here, talking about his own experience.

"It gets thicker and thicker over time, but we have to break through that, and that's hard. In order to carry ourselves well, to break through that, we have to become aware of what's going on in our community as a whole. We have to see what other people are doing.

"And that's the good part! There are a lot of organizations and people working toward these goals; for example, taking care of water. And, there are a lot of volunteer opportunities. And volunteering is a huge aspect of it, but that's a difficult thing too because a lot of people don't want to work for free, yet sometimes we have to see the personal point of applying our-selves and serving this Earth that we live on.

"I don't really know how to dismantle that situation about volunteering or opening up, because you can't really force people to change." He pauses, then adds, "But we're in a position where we have to change because the world's changing so quickly."

Dahr asks, "What do you see as factors getting in the way of our moving in that direction? What are some of the challenges, and how might we find a way to do that anyway?"

"As one example," Steven answers, "the oil industry is a huge organization that blocks the way that we utilize energy and electricity. We use the automobile, plastic, and other things that are dominated by the oil industry on a daily basis, and we need to become aware of that as a whole. There are a lot of people who have made a huge effort to bring that awareness to people, so we have to be able to see those organizations and fight to have alternatives, because there are plenty of them."

He moves from positive work to the impediments, always seeing the challenge while focusing on solutions, on people seeking and forming change. He keeps pointing out both sides of the situation.

"There's a dominant force from the greed, that's taking and taking and has no interest in sharing and keeping things level. There's an interest in collecting and building up the personal pile, and putting oneself on a pedestal. So then it's not equal. That's a huge factor in our big organizations in general.

"Once again, oil is the big one. I see plastic destroying life in the oceans. The gyres are accumulating massive amounts of plastic that then cultivate hydrophobic bacteria. This occurs on plastics of all sizes, and when consumed by marine organisms, these toxins then biomagnify up the food chain. This causes the larger animals to consume a greater load of toxins, bringing upon them severe illness and in many cases, death. With plastics of greater size, we see entanglement of animals and even suffocation from mistaking plastic for food.

"Plastic dominates our modern consumerist lifestyle and is fueled by the Big Oil industry. Many people work to bring alternative sources to plastic, such as hemp and cellulose, which is really incredible, but unfortunately, even those alternatives do not biodegrade because of a polymer coating used to bond the material. We are destroying life on this Earth because of our societal addiction to this extremely useful, yet deadly, resource."

I ask Steven how, as a young man in environmental science, everything he's learning affects him emotionally. I offer that he does not have to go into it personally, but he nods and says he doesn't mind at all.

"At first I was desensitized to it. I felt I could take in a lot of information and that I was just learning about it and it was okay. I'd tell myself, 'I see that.' But as time went on and I took it all in, it began to settle, and I started feeling it, and it ultimately made me very sad. I felt extremely depressed because I saw how much the world and the people and the animals and all living beings are just being exploited. And because of that, everything is suffering. And when you realize that suffering, you take it on yourself. And we all are taking it on. It's not just us individually. We all feel that suffering together, so we have to make the choice of allowing ourselves to really feel it. And it's hard to feel it. It's easy to say we have to feel it, and hard to do it. But we have to. I wanted to say 'No' to it and not feel it, because it's easier and feels better, but the problem with that is there's no growth. There's no room for expanding, and you fall back into isolation."

I ask him if his work with the tribe helps in dealing with the sadness as well as the sense of growth. This is a crucial question for me because I work with young people.

"My work with the tribe has allowed me to be able to feel community, and being able to feel community has been the biggest thing that has helped me feel better about my position, to feel I can keep growing and learning and not let these feelings of sadness totally overwhelm me. It helps alleviate those heavy feelings. Doing the work with the tribe, the actual work, I feel like I'm making a difference every day I go to work.

"A lot of people I go to school with really want to make a big difference in the world. They see all the problems and they want to fix them all. And I always tell them, 'You can only do so much in one day,' because you want to still be happy at the end of the day. You still want to enjoy yourself. And if you try to take on the whole world in one day, you're just going to crumble. It's too much!

"But every little bit helps, one day at a time, and doing the work with the tribe is that. We do a little bit of work, and a little bit more of work, which leads to a lot of work, and that's what helps me not feel depressed. That's what helps me feel like I can keep going, realizing that what I'm doing is actually making a difference. Of course there are steps along the way where you feel like you're taking a step in the wrong direction, but that's part of the process. The path we're on has a lot of twists and turns."

Steven flashes an infectious smile, then Dahr asks, "Can you talk specifically about tribal fire management, both the physical practicality of it, then more about how that specific work affects you?" During this interview, the CZU fire in our community, where Steven lives, is still not contained, already destroying over nine hundred homes and tens of thousands of acres of forest. Dahr's home in Washington is inundated with smoke so dangerous that people cannot go outside.

"Definitely! From what I'm learning, our tribe has traditionally used fire for maintaining landscapes, for thousands of years. We've learned that through archaeological records showing different scorch marks. The tribe burned from intervals of five to seven years up to fifteen years, and we see that as a constant, as a routine burning. The archaeological records tell us this was caused by people, and knowing that helps us realize that we need to be burning as well.

"The archaeological standpoint also helps us identify the plants that were around during those burns, which we then compare to the plants that are in those areas in the present, which are completely different species."

Dahr asks him to say more about controlled burning, because the average person isn't aware of these Native practices.

"There were major areas that were burned regularly, such as the understory of a forest and open grasslands. We'll just use those two examples. In an understory, there's light that comes through the trees and small shrubs pop up. If you observe the forest in the Santa Cruz area, and you look at the understory, it is chock full of shrubs and different vegetation. You can't see through it, but when the Spanish came, they would see eight to twelve very large redwood trees per acre, and they could see through the forest. Now, from the lack of management and regrowth from logging, there are a hundred to two hundred smaller redwood trees per acre, with massive growth of shrubbery in the understory."

Steven talks about the fire danger inherent in this scenario, and how this is a result of laws preventing controlled burning during colonial times, laws that are still in force today. He points out that the laws create fuel for wildfires.

"Because of the outlawing of Native burning, that understory wasn't taken care of. And all the trees were clear-cut, so you now have hundreds of trees growing in a smaller area. Again, when the Spanish first came here, they could see straight through the understory. Can you visualize what it was like? If you think about it in this way, you can see this is what occurs from lack of management.

"Also, because it was burned so routinely, it wasn't an intense burn but was a really light burn. But now if that vegetation catches fire, it's out of control because it's such a heavy fuel load. That's the key! You have to utilize fire in order to prevent massive fires."

Steven describes the situation, then once again quickly goes to solutions. "It's difficult in California with the laws, but there are certain counties that allow burning. In Santa Cruz County, we work with California State Parks and CAL FIRE on controlled burns, and on fuel reduction in general."

Very attentive to Steven's points, Dahr brings up where he lives. "We just got out of some of the worst smoke in the world from wildfire smoke from California and Oregon. That's the context of our conversation. In the work that your tribe has been allowed to do, are you seeing those areas as safer now or in a better position relative to the fire season?"

"We're hoping to find the answer as soon as possible, but we don't have it quite yet. We are too early into the project. We haven't done the project in its entirety to really show that data, but the area that we worked did not burn nearly as heavily as the surrounding areas. So we have a little example of where our work compares in this fire."

Steven leans forward to add what must be said, as he knows the tremendous damage this current CZU fire is doing to humans, animals, and the forest. "This fire that's happening is tragic! It is such a shame, and unfortunately, is a result of lack of management. This is a result of people, not Nature."

He describes how some people see fire as Nature being wild, "crazy" even, but he protests this view adamantly. "This is a result of *man*, not Nature!"

He elaborates on how many people say wildfires are normal, but "they shouldn't be this kind of normal. It's not supposed to be this intense. The whole state is on fire! And as I said, there is no management because of the laws that are in place." He pauses, then declares emphatically, "These are laws based on the concept that man is *separate* from Nature!"

He comes back again to the idea "that humans can operate with the tools of Nature for the benefit of Nature. This older concept is more interactive. Relationship to Nature rather than removal from it."

In light of this comment about relationship, I ask Steven what the fruits of traditional burning are for him personally, since he has talked about confusion, community, isolation, relationship, and finding purpose.

"The controlled burning is coming from an ecological standpoint, which is crucial and important, and the cultural aspect of it is coming from a relationship standpoint. In having that relationship, we realize that what we're doing is caring for those creatures and plants. We are aiding them in their growth and their ability to keep living. Even though fire may appear to be destructive, it's not all destruction. It's a very loving force.

"This ultimately brings it back to what I said before. This work makes me feel like I have purpose. It makes me feel fulfilled. What we're doing is giving, so it's revitalizing. It's a breath of new life through what is observed as destruction. We can observe cleaning our houses, and if we clean our house, then it feels good, everything feels great and you can breathe.

"And that's reflective of my life, of having a new start and coming into the position of working as a steward. Whenever I work on fire, I'm connecting with that very land I'm standing on and working with. And I feel like it's a prayer rather than a force of destruction."

I ask Steven what grows to be obvious from his words: "When you work, do you start the workday with prayer?"

"Yes. How the workday goes, and in general for the whole project, is that we pray for the trees that we cut down; we offer tobacco as an exchange for taking that life. In our prayer, we acknowledge our love and care for that individual tree. And we also acknowledge that we are here in a good way to steward this land and to let that tree know that 'we're only removing you from here because this isn't where you're supposed to be, but we will take care of you and your relatives, where they are and where they belong.' We make sure it's clear that we're there to take care of this area. And that we will take care of their relatives in the area that they are naturally accustomed to.

"When it comes to fire, we're working toward being able to start our burns with the traditional fire hand drill. And that's really important, even though we're using modern tools alongside that process. We're initially starting it in the same way our ancestors did."

I ask how the tribal approach sits with the people from CAL FIRE, and Steven brightens up, very pleased to answer.

"It's actually quite fascinating. You see a side of people, who have been doing this work for years, come out, and it's teaching them something new, something they've never seen before. It's foreign to them. And it's really powerful because I've seen nothing but good support coming from those who are with CAL FIRE and State Parks. They are open to it, and they're open to listening to us and our prayer of getting ready for a burn, or getting ready for a day of cutting trees. They want to be a part of it. And it's fascinating how well they take to it. There's no opposition whatsoever there. They ask us questions, which is really cool. It's a blessing.

"About my other Mutsun relatives that I work with, and how they feel about it, I can't say how they personally feel, but I can only imagine that they feel like I do. All of us go calm, and we get really silent. We go inward when we're burning, reverent of the fire that is taking place. We are letting it be what it is, allowing the fire to happen, while realizing that we are responsible for the fire at hand. Not letting that be something with fear, but with a good, strong mind and strong heart. That's important, to mix the two when burning!"

Dahr is moved by Steven's statement: "That's beautiful. Thank you. Are there any other ways you'd like talk about where ceremony fits into the work that you're doing?"

"Ceremony is life itself. Every day is ceremony and every day we're learning and praying and being in direct connection with Creator and our ancestors and our world around us."

He pauses, then tells us he wants to talk about other specific practices he's involved in. "We have two ceremonies that I'm part of. One is the talking circle in our stewardship practice, especially when we have the internships, where we take in tribal youth who come work with us.

"My group took on two younger women interns, one who just graduated from high school, and the other in the middle of high school. We teach them about the work we're doing, and we have the talking circle weekly when the youth are there, with Val Lopez. We open it up for the tribe to talk from their hearts, so the youth and the stewards talk about their experiences and create an open sense of a circle between us, where we can share everything together.

"We've also been doing sweat lodge with Shannon Rivers, in the Lakota way. We don't get our tribe's sweat lodge, but we still have sweat lodge, and our members share our songs. We have women in our tribe who know quite a few songs from our tribal history and tradition that are shared in the sweat lodge. They're very powerful songs, and the Lakota songs are very, very powerful too.

"It could just be from my ancestry, but whenever I hear our tribal songs, they connect easily and deeply, and, hearing those songs, it helps with the sweat lodge process and prayer of allowing myself to go deeper, and allowing me to feel deeply. Hearing those songs that have been carried on through our tribe for thousands of years, helps heal us today."

There is a long pause, and, hearing Steven talk about the value of healing in ceremony, I think about the Amah Mutsun's current struggle over Juristac, an ancient ceremonial ground in his people's homeland, used for millennia by the neighboring peoples, slated to be destroyed and made into a quarry. The Amah Mutsun are not a federally recognized land-based tribe, and they have no treaties to protect or acknowledge their tenure and right over any of the land and its stewardship.

I ask Steven to comment on Juristac, stewardship, and the potential of solidarity between Indigenous and non-Indigenous people.

"What's happening with Juristac is tragic, thinking of the intentions the developers have to manipulate the land. It was historically a ceremonial ground for our tribe and many other tribes, and the fact that it's a ceremonial ground means that it still holds those prayers. It still holds ceremony, in the land itself, and to manipulate the land, you remove that prayer from the land. The prayer is still there, but 'development' today diminishes what has been integrated into that land for thousands of years."

Steven stops and contemplates, visibly upset, then apologizes for losing his track. I think of the thousands of years of tenure on that specific place, and how the Native people today have absolutely no rights over it. He questions whether he's answering my question. Very moved, Dahr assures him, "You are addressing it, absolutely, very directly."

Steven nods, then wants to say more, which is the most powerful point possible. "It's sacred land!

"And, I also see the project from having done construction work in my past, so I know concrete is a very valuable resource that has helped a lot of projects, and it's an amazing formation of materials. But what's going on is not right!" He is emphatic. "It's sacred land!"

There is a silence between us all, then Steven talks philosophically about the settlers' desire to make things permanent and stable, like the laws based on humans being apart from Nature, using concrete as a metaphor for an approach to life that's gotten us into deep trouble. "We want to make things steady, to stay put, but historically nothing really stays put. Even rocks themselves are in constant motion. Everything's moving all the time and permanence is imaginary. We just think it's permanent because it stays where we want it for a certain amount of time that we can witness. But over time that concrete's going to decompose." He comes back to today's monetary priorities being willing to destroy a ceremonial ground thousands of years old. Thousands of years.

"What really weighs heavily on me is the question of where those people are coming from. Why don't they see the other side, that it's a deep, deep place rich in a living history? Healing and ceremony has taken place on that land! For someone to not see that, and to come in from a greedy stand-point, is one of the illnesses that man is facing. All they see is profit. They are not allowing themselves to see the importance of the land they're taking."

Steven catches fire.

"People have been taking from the Native people since they first showed up here, and this is the continuation of that colonization. It is that very problem that has been facing this country. It is stealing from Native people, and stealing from people in general. Like stealing from the impoverished, it is an ongoing problem. It keeps happening.

"To be able to fight that and succeed, and really allow Juristac to go back into our hands and management, not into our ownership, but into our be-ing able to take the responsibility for that land's health once more, would not only help the land and our people, but all the surrounding communi-ties as well."

Wedding the philosophical, moral, and practical aspects of his people's situation and how they fit into all people's necessities, Steven points out the many effects the mining will have, from damaging the water table, to the quality of the groundwater for surrounding towns, to endangered spe-cies living there. "It affects everyone," he says.

I ask if the local surrounding areas are in support of the Amah Mutsun efforts because of this.

"We have a lot of positive support from certain counties. We're finding some support through government organizations, but we're finding a *lot* of support from the actual people who live in these areas. That shows the aspect of community I was talking about, those who are seeing that we're all in this together. This is a Native issue, but at the same time, it isn't; it's a community issue. It's a world issue because this is a result of that con-tinuing of taking and taking. To fight that pushes us further away from iso-lation and brings us closer to community."

Steven's words bring me back to his descriptions of starting the workday with prayer, with CAL FIRE and State Parks members participating. Because youth are the future, I ask Steven if he has any words for young people, or for all people, but from his standpoint of being young.

"Realizing that I'm only twenty-four, I feel like it's about honesty and being truthful with yourself and those around you. And yes, it's easier said than done. It takes a lot of courage. Like I said, you have to allow yourself to be vulnerable in order to be strong.

"If you are vulnerable, you are able to see it from the eyes of a child of innocence. Vulnerability is key, but vulnerability is tough because you leave yourself open, and when you're open, you take in a lot. And it's hard. When you take in a lot, it's hard to cope, and you have to take time to do that.

"I have made it a key in my life to open the door and look at the world for what it is. I don't want to be lied to, and I don't want to lie. There is so much to decipher, and if you really allow yourself to be patient with your thoughts and your feelings and to observe them, and not act impulsively on them, you'll be able to start seeing and feeling truth.

"It comes out very clearly at times, but at other times it's not so clear. And as my mom told me, when we don't see something we want to see right away, such as truth, that means we're not ready to see it right then. We have to allow ourselves to not know the truth in that moment. We have to be okay with not knowing in that moment. It's really crucial. You have to take every day to remind yourself of this. And that's hard.

"With climate change, both on the emotional and spiritual side of it, what you really feel is what you allow to come through. But there's a physical aspect of it too, and they all go hand in hand. You have to have an open sense from your heart, from your mind, and from your spirit, and your body. And your body has to act, and that's where action comes into play.

"You want to be active, but you don't want to jump into action too quickly because sometimes you can make the wrong decision. Sometimes you have to strategize. You have to look and ask, 'What is the best possible way to act through this?'

"If you're patient, you find a niche. The way the world works, and the way biology works, is that there's always a niche. There's always a place where something fits, and it may not be clear at first, so you have to allow yourself to not know in the moment, in order to really find your niche. I found my niche through the Land Trust and doing hard physical steward-ship work. For others, it may not be the same. And that's okay, because as a collective, we can make this go in a better direction. And everyone has their part."

There is an air of finality in the open nature of his closing comments, and we ask Steven if he has any final thoughts to add. He becomes very deliberate, choosing his words carefully.

"I heard someone say recently: 'We are the descendants of genocide.'" He waits, letting the words take root in the air.

"It's important to realize that, and though it comes from a Native stand-point, many other people are descendants of genocide as well. Everyone is, in one way or the other. It's important to realize that. It's crucial."

He pauses again, letting those thoughts remain, then reflects on another thought, not connecting the two overtly, but allowing the transition and the movement of connecting ideas, one idea moving to another, all infer-ring the choices found in how we relate to where we are now, to how we got here, and to how we proceed.

"The sun's fire is the energy of life," he finally says. "I think about how the sun gives us energy, how it gives us life. This is what fire can do to the landscape. It helps renew life, and bring back life. Fire isn't destruction alone. It's a bringer of new life, rather than of death."

9

Marita Hacker
(*Hunkpapa*, Norwegian)

Change

COMPOSED BY DAHR JAMAIL

I am not depressed or scared or have a lack of faith. Because for me, there's been enough to show me, *this* isn't permanent, this *state*, this *condition* isn't permanent. It will get better in some areas, and maybe worse in others before it can get better. It's cyclical. It's a cycle. It's a circle. It's a spiral. It's never ending, ever changing.

—Marita Hacker

Marita Hacker is an elder of Dakota Hunkpapa and Norwegian descent. One of just a few female Apostolic bishops in the United States, she has also been involved with hospice nearly three decades, and it is her passion.

Marita is a longtime friend of Stan and, for a few years, me as well. She lives in a remote village in southern New Mexico, serving her community, but also those who travel to see her and her husband for ceremony, other spiritual work, and friendship.

Marita spoke with us from the front porch of her humble adobe home, surrounded by the rolling hills and mountains of Apacheria.

After talking about the climate catastrophe and all the other converging crises, we began by asking Marita, "How did we get here?"

Marita replied, "1492," and we all laughed. She went on to say that she thinks it happened when people lost their connection to "that which is greater than us."

Marita explained how this is due to, as she sees it, the human population explosion coupled with growing intolerance for differences, exacerbated by people's wants eclipsing their needs, and the ensuing lack of satisfaction with "what is."

"Nobody wants to be without. When that happens your spirit is focused only on survival. It's not focused on what it can do, what it can bring, what can happen."

She spoke of how Earth has changed many, many times, and talked of us going into the fifth world now. "Earth is a self-sustaining entity that is, right now, seeming to cleanse itself of that which irritates it in a variety of ways," she said, before going on to point to several wildfires and hurricanes that raged as we spoke, in addition to the pandemic.

"She's cleaning now as we speak," Marita said of all the ongoing natural disasters. "It's horrific, and it's not going to change until we get back to that grouping together, to being caring, because we can't change the world; all we can do is change ourselves."

Marita thinks we can still influence the changes that can happen globally, but that this influence must come from each individual.

"Individuals, creatures, and everything that encompasses the occupancy of this planet is a puzzle piece. And the puzzle piece that I have is no greater or less than you are, Stan, and no greater or less than Dahr is, or the horse across the road that I'm looking at. We are all pieces of this intricate puzzle."

Marita spoke of what dismayed her about what she currently saw happening amongst the puzzle pieces that comprise life on Earth.

"Unfortunately, some of the pieces have been lost, gouged out, spread apart, lost, and somehow it needs to come back to cohesion."

Her sadness was palpable as she spoke about the crises the planet was suffering, and at times her voice wavered with emotion.

"How do I keep from being depressed and disgusted? The sorrow is all encompassing, but all I can do is get myself out of it."

She paused, looked out at the mountains while taking a deep breath, then continued.

"What did my elders give me to hold on to, *especially* for this time? I'd be bat-shit crazy if I hadn't had the curtains opened to show that there is so much more than having tangible objects that don't satisfy that inner self, and they never will.

"It's the awareness that it's in the consciousness of humankind that needs to step up, *a few levels*. It doesn't matter if you're Indigenous, if you're full blood, if you're mixed blood, if you're Asian, if you're Black, if you're Lebanese, if you're Islamic; it doesn't matter. You're part of a puzzle piece that is intricate to this world."

She thought for a moment, then added, "All this intolerance of others is a reflection of the intolerance for the planet."

Marita was impatient, given the gravity of the converging crises. She wished that the flashes of knowing and feeling of depth that occur when she is in ceremony could happen to everybody, all at once, to have an immediate shift and change, and then the ability to come together as a stronger whole to go forth.

"Many beings that live in this day and time are damaged, whether it's personally in their family system, whether it's culturally, or whether it's a reaction to another culture attacking it. So everybody holds on so bloody tight, to *something* that is tangible, but it's not tangible. They do that, instead of focusing on the healing process, and on the acceptance of this as being the short time we're here," she said. Her decades of hospice experience shone, as she repeated, "A *very* short time."

If Marita had her way, each of us would pause, and ask ourselves what is most important.

"What is your passion? What's your passion that doesn't harm another being on this planet? So much of the passion has been taken away."

Instead of people living in their true passions, Marita sees "groups of colliding energies" not doing anything truly productive, but instead grabbing and accumulating and saving material things. She pointed to the run on toilet paper in the United States during the initial stage of the global pandemic as an example.

"The toilet paper thing is over, but what else are they grabbing?" she asked with a smile. "Is it back to making money for your second cabin? Is it drilling for more oil? Is it fracking and causing a gazillion earthquakes in Oklahoma? *That's* your *passion?* There's something wrong with that."

But she has a deep faith, belief, and as she put it, "an intrinsic knowingness" that where we are now will change, because it always does.

"There are ups and there are downs and there always will be. I can't affect what is going on out there, but I can affect this," she said, then pointed to her heart. "I can affect me. And in affecting me, whoever comes into my sphere, they can have an exchange that's positive, not negative."

After a long pause, where the three of us took some long breaths, I asked Marita if she would speak to the importance of ceremony during this time.

"It's *very* important. My elders were so emphatic about this. Henry Tyler, Martin Highbear, the strong women; what they taught me . . . going back to the tangible and intangible, is that we're actually kind of intangible. We are this swirling mass of atomic energy that they can't find under a microscope, as finite as they attempt to get it. We are a combination of movement and the light. The English term they have for this is energy. We are all units of energy. We are puzzle pieces, and there is this battery that is an unlimited source of energy that we are connected to *all the time.*

"But most of us don't have the constant awareness of this deep, perpetual connection to it. And that is where ceremony comes in. We are all busy with *different* kinds of ceremony, that which comprises the tasks of our everyday lives, from going to work to hunting a deer to raising kids."

Marita said the difference is that when people come together as a gathering of like-minded individuals and focus their heart process onto a theme, as it happens in ceremony, they are tapping into "that battery, and more energy occurs, and then you have results."

Examples of the "results" she gave are things like being in relationship to rain, healing illnesses that maybe Western medicine had written someone off from dying of, or "you have the ability to drive an automobile without putting a key in the ignition, albeit for a limited time," she said, chuckling. "To me that is ceremony, and in the ceremony, magic, for lack of a better term, occurs. And when that magic affects more than one individual, *they* are changed. Their inner being is changed, because they have experiential knowledge of a different type of living in a different type of reality. And *that's* why ceremony is so important."

Marita repeats that she sees humanity getting caught up in the minutiae of daily life, and when that happens, we are forgetting that we are attached to "the battery" every minute.

"The connection with the battery is what makes everything vibrate. Now, how do you change your vibration? And what affects your vibration?"

She explained how lower forms of vibration we surround ourselves with create lower vibration within us, as well as the opposite happening with positive things. "The strongest positive repels the weakest negative; however, the negative still exists, and it's not bad. It's not good. It exists also. It's 'What do you choose?' It's 'Where do you choose to put your consciousness?' Is the world going to shit? Well, kind of. Does it have its beautiful places?"

She turned her phone around and showed us a beautiful landscape of hills, trees, a riverbed stretching out in front of her home.

"Look at this, look out your window," she said. "We *are* in paradise. We are in *Eden*. *We* didn't get kicked out. Our *minds* got kicked out. The mind became the all-powerful thing. The Tibetans say, 'Take your mind, put it in a box, and enjoy the day.' And when you do that, you have a different experience of reality."

Marita believes that, globally, people are aware of the great changes that are upon us now. Referencing the five stages of grief (denial, bargaining, anger, depression, acceptance), she said the current country that is the

United States has been in the denial stage for a long time. She thinks the country, generally, has been childishly blaming others for its own actions, not taking any responsibility, even with power, hence the corruption due to its lack of consciousness. In this way, the power feeds into those who wield it out of their desires to fill their inner lack. She sees the country repeating the same dysfunctional pattern, and it baffles her.

Similarly, she is unable to understand war, as the idea of two human beings attempting to kill one another is beyond her comprehension. Until humans stop seeking war, the imbalance in the Earth system will continue, and she sees that as having gotten us to where we are today.

Marita also suggested that the lack of a rite of passage for men played a role in setting the stage for this, so I asked her if she would talk more about that.

"Well, here's the result," she said, laughing. "There isn't any rite of passage. We're not even a nuclear family system anymore." She discussed growing up in Germany, and how her mother's family had their farm taken away by Hitler, but before this all of the related families lived close together, oftentimes sharing a home, with everyone regularly going to one another's homes, and how this wasn't much different in the United States until after World War II.

With the advent of the nuclear family and lack of connection to relatives around that time, the absence of the rite of passage was felt more deeply, as Marita sees it. For many Native communities, that family has been very deliberately broken in order to weaken them.

"In order for a rite of passage to occur, you have to have a family system that acknowledges the wisdom that is being obtained," she explained. Marita cited some young men near her home who work the farm, and work with horses.

"They are taught horsemanship, respect, responsibility. Then a rite of passage for them is to prove an example of their horsemanship abilities, or to rope a calf, or find a calf that's lost. What I find interesting is that the whole community fosters those kids' first ride, or first contest . . . and then they're men. And they are compassionate men after they go through that. They're not going to shoot a rattlesnake for the hell of it. They're not going to go hunt them in the wild just because they're rattlesnakes. They will only kill one if they have to, if it's threatening their family. That's a rite of passage that's actually happening in front of my very eyes with a different culture of people. And I love those children. They come here, they visit and ask me how I'm doing, and ask me if I need anything."

She doesn't see many of the other kids behaving this way, because their parents gave them to babysitters and didn't foster them, or acknowledge levels they achieve as human beings. "And in that, you can grow up wondering if you are doing enough," she said, "wondering if you are doing okay, asking yourself what you are supposed to be doing, how you can effect a change. How are those kids going to know to listen to that still, soft voice inside of themselves that enables them to survive?"

Stan attended two of my readings during the tour for my book *The End of Ice*. We both noted a widespread expression of helplessness in response to the book, because the information in it is so overwhelming. It was common to hear people say, "It's so overwhelming," then shrug and conclude, "Nothing I do could make any difference whatsoever." This often led to people thinking, "Well, humans are just messed up and so the Earth will be better off without us."

"There was no sense that we're the children of the Earth," Stan told me at one point.

This is what Marita was tapping into, this same feeling of aloneness that so many people feel, and how that tied into what she was saying about how "we are in the garden, this *is* Eden, we weren't kicked out."

Marita saw that as another form of denial, another way of pulling the proverbial covers over one's head, and was sympathetic to those who felt the need to go there, given the gravity of the crises. "But don't stay there," she said. "Their minds are turned on, since science has made the mind the end-all. But the mind is not a unit that operates independently without the heart. We have to shut off our minds, and find what we can do as an individual."

I asked Marita if she would fill in the story of the rest of her life for us, because the trajectory of her life is how she's chosen to respond to everything going on.

"My path was my destiny," she said, before going on to say that there was already a formula in the Earth and the Sky for her. "Without that, I don't think I would've gotten here to this point, this place."

Marita's mother was Norwegian and was forced to relocate to Germany, thanks to Hitler. Her father was Hunkpapa Sioux, later joined the U.S. military to earn a living, and ended up in West Germany, where he met her mother. Before Marita was born, her father died in an accident, but even before she was born he was clear that he never wanted her to go to a reservation, instead wanting Marita to be raised in a different culture, and to go to college. He also did not want her to go to the United States.

So, naturally, when Marita turned twenty-one she moved to California in the middle of winter, finding a happy respite from the winter gloom of Germany. Various jobs kept her busy. She went to a reservation along the way, but was unable to find a sense of family there, instead seeing what she'd spoken of: aloneness, lack of like minds, and the feeling of separation.

"It can be devastating, heart wrenching, or you can take it and utilize all of it and make a path that works for you, which is exactly what I did. I wanted a child, so I got one. I should have asked for a better husband, but eventually I got one of those too," she added, laughing. "I had to go through everything else to understand what I needed and what I wanted for the rest of my life. And it happened. Now Creator is first in my life, then anybody who is hurting, then my husband and my family."

After what she described as an "intense Sun Dance ceremony," Marita left California for Oregon, where she lived off-grid for seven years, also spending time on a Hopi reservation, and in the homes of different Native folks as she "let them foster what I had within me."

She spent a good amount of time in the Lost Creek Lake area of the Oregon wilderness with an Apache elder who told her she had to leave the state. She didn't accept this at first, happy living in a place with a lot of human contact. "I loved it there," she said. "I was so very content. To be informed I must leave to go to a more populated area was akin to banishment and very, very painful."

So she was faced with the choice between sticking with that, or going "deeper into the wilderness. He told me my knowledge needs to be shared and given to others, 'of your own kind, as well as with others not of your own kind,' he said, as 'you don't have a specific people that you're going to minister to.'"

Marita had moved to California, but, going against what her elder had told her, attempted to return to Oregon. In the process, her car was totaled by a drunk driver. She shared what happened when she called him to tell him what happened.

"He said, 'You were trying to come back here, weren't you,'" she said, laughing. "I asked him what I was supposed to do, and he told me, 'The stars will guide you, my daughter.' And sure enough, they did."

Her first sweat lodge was at a Sun Dance. There, she was handed a fan by the intercessor who told her, "Tag, you're it. It's time." Her job was to dust the women coming out of the Moon Lodge. From there it went to helping prepare her brothers and sisters for Sun Dance.

"I needed a man in my life, someone who could understand this path, and my husband showed up and hasn't left yet," she said with a smile. Marita said they share the task of working with men and women, and, as she put it, "We play off one another." She smiled while she reflected on this for a moment. "All this just unfolded."

She believes she was meant to become a minister, a priest, and she always wanted to be with people who were in hospice. "I wanted to be with someone who was dying because I knew I can effect less suffering in that area of their life. And in order to do that, I had to become a minister."

For her, finding a Christian faith was difficult. "I was trying to enter into a patriarchal stronghold," she said.

She's been doing hospice work now for nearly thirty years. In order for her to accomplish this, Marita acquired a doctorate in divinity. Part of her

ongoing practice is assisting others in finding their own divinity, and helping them strengthen it.

"How do you become the power that you are intended to be," she stated
as being the operative question.

After doing hospice work in the Monterey Bay area for decades, including directing a program and being a chaplain, Marita moved to her husband's homeland in Apacheria, a long-standing dream for them, where she
took a position teaching in a local high school. However, she eventually
had to step out of her high school job as a teacher when she became gravely
ill. After a long, slow recovery process, she decided not to return to teaching. I asked her why.

"I disagreed with the administration about what they were trying to teach
the children. And I disagreed with how they were teaching the children."
The restrictions on the human imagination ran counter to what was necessary in effective hospice work, in helping people "walk on."

Despite that, Marita had taken free range in how she taught. "I had young
girls doing telepathy," she said, laughing. The girls were intuiting each other's
thoughts without cell phones. "They've never been taught anything like that.
They're good little Catholic girls, right? I showed them something different
than a book. I showed them their own ability to tap into another reality."

The kids were overjoyed by the experience, and beyond enthusiastic.

She described the U.S. educational system as "a cookie-cutter formula
that isn't working," and that she had grown weary of fighting the system
that did not foster the kids within it. She saw total and complete apathy
within many of the kids as the result of this, and how it wasn't until kids
got out of that system that she saw them find their passion and follow it.

Knowing Marita's own work as an intercessor and chaplain over the decades, we asked her if she would speak to the power of prayer in dealing
with these issues.

"It's energy; the process of it is an energy shifting exercise, as is meditation," she said. "When you end up in that plea, so to speak, you shift your

energy. The *manner* of the prayer is also key. Sorrow can be sublimated by meditating. It can be taken away just by focusing on your breath. So can prayer by repetition, and chanting, which shifts the spinning of our molecular structure, which goes out into the universe and never ends, and bounces off all the other molecular structures."

She believes sorrow exists so that we don't repeat what caused it in the first place, and it can also be used to mitigate suffering for someone else.

"If people have the same prayer, it just might come true, because that has been proven repeatedly," she said. "That magic is what keeps me going. The end result of this experience is that it *works*."

She shared a key example of this by way of her hospice work. Speaking from her decades of experience in that field, Marita mentioned how many people, as they near their death, experience a review of their life, and many have regrets of not doing something, or from overdoing something else.

If they are able to express what that is, Marita works with them to try to find something they can achieve in their last moments of consciousness.

"It's magical, beautiful, wonderful, amazing, and very, very powerful. In that, even though they leave their husk, their molecules and energy go on forever, bouncing against whoever. So if you lived every day as if tomorrow you may not be here, you would live differently. You would search for joy. You would search for passion. At the first sign of bitchiness or grumbling you turn to shift it. You'd ask, 'Why is this happening? You're not the problem, *I am*.' Then *you* take the reins and change *yourself*."

I asked Marita to talk about her having been welcomed into the Order of Templars.

"It was about Spirit, and about medicine. I was *so* pleased when I got my certificate from the Church of England, when I was welcomed into the Order of Templars."

Marita decided to become an ordained Templar because she sees them as a very cultured group of people, of varied walks of life working to become better spiritual beings, from bankers to Catholic priests that belong to the Order, to people like herself, "working in the background with ceremony, to effect change on the planet," as she put it. She also took the step as it allowed her to serve a broader range of people in hospice, as well as the fact that it gave her the ability to give people their "last rites."

She describes the work of the Templars as similar to what happens in a sweat lodge.

"When everybody gets into the lodge there's a focus . . . everybody's on, and you can effect change," she said, reiterating what she'd mentioned earlier. In being part of the Templars, in order for this to happen for her, she was allowed her own form of worship, in accordance with her walk of life. She describes herself as a spiritual warrior, "and in that comes the magic of medicine."

Her proving ground for joining the Order of Templars came in the weather.

"You have to either make it rain or stop the rain, or flooding, or a hurricane," she explained. Marita said she is better in working with wind. "My ability to play with the wind is greater than with any of the other elements," she said. Instead, however, she was asked to make it rain.

Someone from the Order tracked her, while she set up her altar and prepared herself for the action. There would have to be a meteorological reading to prove she made it rain.

But just before she began, she received a phone call from a young man on the San Carlos Reservation, whose brother was locked in a room and threatening to kill himself.

"I'm all ready to do my proving and I get this phone call about the suicide," she said with her eyes widened. "I decided to do both at the same time because I really, really, really wanted to be an ordained Templar and I didn't want to fail, so I had to do both.

"So I go in and I do this thing and it rained on this small little area in San Carlos where this man and his brother were, and it was measured meteorologically. The rest of the rez didn't get any rain, just that one little corner. And the young man was able to experience me, and is clean and sober to this day."

When this information came back to the Egregore, Marita had to explain what happened, because they could feel her distraction and deliberated as to whether to allow her into the Templars. They did, but with the instruction for her not to split ceremony again, if she could help it.

We talked about how one of our hopes for the book is to provide perspective to help people emerge from the fear that is so ubiquitous around the planet at this time. We asked Marita what she would say to folks who are stuck in that fear, or in a state of panic around the converging crises.

"Those people in fear have lived on this planet long enough to know that change is a guarantee," she said sternly. "And they have to stop being afraid of what hasn't come in yet. And if they're faced with something that is fearful, they must ask themselves, 'What can I do?'"

She used COVID-19 as an example, pointing to how there are many sides to even the most horrific things.

"I actually think it *can* be seen as a good thing, because it showed people fear, right there. Then it showed greed, number two. It showed a lot of things to many individuals all at the same time, if they went into it deeper."

Marita talked about "this time of separation" that we are in, reminding us how tribally, in the Indigenous world, people are sent out to seek a vision to discover more information about themselves.

But she sees the United States as a country that has thought it hasn't been able to take that kind of spiritual action, to look inside. But thanks to the global pandemic, the entire planet has had to do so.

She gains faith from the fact that the number of people globally now taking the cue from the virus and looking inside is causing a dramatic increase in awareness.

"By where they live, by what their individual issues are, the strong ones are going to come out of it with the answers; the strong ones are going to come out of it with the ability to effect a change in their circle which will effect change we can't see yet. And it's gotta be positive. And it's got to be negative too, in a cyclical pattern because that's the way the universe is set up. We have days of rain. We have days of drought. We have days of flood. We have days of hurricane. We have days of Nature being quite peaceful and beautiful. You know, it is an ever-changing system."

Marita pointed to how, misguided or not, people are now exercising their voice, "Be it with fear, or panic, or with an inability for tolerance.

"We're in the middle of all of it. And we were told we would be. And if people don't have prayer, what is their release? They don't have to pray. They can look at their grandchild and feel a jump-start in an energetic way. That's sufficient. They can look at a pretty bird, and say, 'Wow!' Use *that*, if that would make you feel connected. Our ideas of God have shifted in the last forty years, and now it's clearer than ever. She quoted Rumi, 'There are a thousand ways to kneel and kiss the ground.'"

She paused, then returned to talking about spirit.

"We need to come back to the belief in magic," she said, using the words "magic" and "spirit" synonymously. "My dream is that people get to have some kind of experience of that to give them something to hold on to. I do. I wish that for everybody I see or touch, because it's life changing."

It matters not to Marita how "right- or left-brained" a person you are, she simply implores us to remain open to spirit.

"We're lucky," she said, referencing people who are practicing the old Indigenous ways. "We're lucky we have those experiences over, and over, and over again."

Of those unwilling to be open to this, she asks "What can you lose? Why are you holding on so tightly?"

Marita, her husband, and Stan used to live nearby one another when they lived in California. Marita shared a story from that time, to illustrate her points about the importance of ceremony, as well as close community.

Stan was experiencing severe pain and ended up in the emergency room. He called Marita and her husband, telling them he wasn't going to allow the physicians to do anything until they arrived, because the physicians didn't yet know what was causing his agony.

When Stan rang them, he asked specifically for Marita to come, not just her husband.

Stan explained how, when they arrived at the hospital, they were told only one person could come into the room and they had to be immediate family. In all that pain, he had to threaten the hospital and the doctors with a lawsuit for denying his religious freedom in order for them to be allowed in to see him.

They both laughed, then Marita continued with the story.

"We get there, and I see him lying on the table, and I left my body, and this other Marita strode over to the table beside the nurses and the snooty doctor, and I laid my hands on him and I looked up at them, and I looked at Stan, and I said, "'Kidney stone.'"

Marita looked into Stan's eyes, and said, "You know how to do this, so let's do it!" Her husband began singing a protection song while standing at Stan's feet, and Marita's prayer lifted him up out of the pain in his body, and all his pain vanished.

The nurse saw what happened. After Marita and her husband left, she said she'd never seen anything like that, commenting on how beautiful of a thing it was to witness.

It did turn out to be a kidney stone.

After this Stan was taken to a kidney center in Los Gatos where he was placed in a bathtub, knocked out, and sound waves were used to break up the massive kidney stone. The technique created tiny pieces that he would be able to pass, but there remained a piece that was too big.

Later, in pain from that piece, Stan came to their house, and as Marita put it, "I actually pulled that stone into a channel where he could release it, which he did when we finished our song." They put Stan on a blanket, again with her husband singing a protection song at his feet, and Marita kneeled nearby, leaning over Stan.

She put a stone on Stan's body over that piece of kidney stone, and just before she began singing into the stone and moving it down the tube, she said, "We've been dealing with these stones for a very long time, living through times and places with very little water."

She sang into it, made a tone into it, and then she moved it. He could feel it moving with the sound and with the movement of the stone on his body. Then it went all the way through and he went into the bathroom and passed it out into a cup, and it was the size of a ladybug. He took it to the urologist the next day and the doctor said, 'Oh my God, how did you get this through?' He told her it was done with sound and with song. The doctor nodded and smiled, paused, then said, "Well, it sounds like our medicine is moving in the right direction, because we're using sound now too."

Of this experience, Marita was emphatic.

"That kind of thing is *real*," she said. "It's very, very *real*. But most people don't get to experience that. This actually happened. This is a true story that all three of us were part of. And there's more, there's plenty more. It's about the ability to step into your magic, your own flow, your own change of energy that can effect huge, huge change, and affect other people who observe it."

Marita paused and took a couple of long, audible breaths.

"That is why I am not depressed or scared or have a lack of faith. Because for me, there's been enough to show me, *this* isn't permanent, this *state*, this *condition* isn't permanent. It will get better in some areas, and maybe worse in others before it can get better. It's cyclical. It's a cycle. It's a circle. It's a spiral. It's never ending, ever changing." Her voice rose on how life is always "changing."

She took a long look out at the mountains visible from her porch, staring at the horizon, then looked back at Stan and me.

"And *we* should too," she said. "*We* should change too."

10

Shannon Rivers (*Akimel O'otham*)
Balance

COMPOSED BY DAHR JAMAIL

Justice probably won't happen in my lifetime. It is a slow process, right? Justice through the lens of Indigenous peoples has many layers; it is rooted in a moral compass, an idea of balance, an idea of relationship building and interconnectedness between the Earth and our love for our fellow two-legged and four-legged and winged relatives. That's the moral compass that I believe that we bring into the prison systems, and that we bring to and from the reservations. I believe that we can bring balance to people around the world.

—Shannon Rivers

When Shannon Rivers spoke, the deep tone and steadiness of his voice, coupled with the breadth of his analytical and temporal perceptions, elicited my full attention.

Stan and I had spoken with Shannon before to give him an idea of how we would approach the interview, during which the consistency and balance evident in his life, work, and voice was obvious. Talking with him felt like leaning against a large tree.

Stan asked Shannon if he might address one of his areas of expertise, that of the role of the climate crisis in causing people to leave their homelands, and the impact upon them for doing so. Wearing black-framed glasses, with his black and graying hair pulled back tightly in a ponytail, Shannon began by thanking us, and with a big smile said, "Good to see you guys." After a pause, he began.

"I have to start with my mom and my father."

Shannon's father was a Korean War veteran, a stepfather who'd come into his life when he was about six or seven years old. He was a large, tall man, and Shannon recalled the hump on his father's shoulder from when, as a paratrooper jumping behind enemy lines, he was shot in his collarbone, which shattered. The injury never healed correctly.

"He had this big, old thing right here," Shannon said, pointing to one of his shoulders. "He would walk around the house with no shirt on and he was kind of scary to me. He was a drinker, and he would drink every day. He would get drunk almost every night, and then he'd go to work."

Shannon described how his father would drink with his friends late into the night, then awaken at 3 a.m. and go to work in the cotton fields amidst the stifling Arizona heat, where temperatures regularly hit 110, or even 120 degrees. Working seven days a week, his skin darkened by the constant exposure to the sun, his father had been given the chance to relocate as part of the 1950s relocation program.

In the early 1950s and extending into the 1970s, the federal government attempted to "get out of the Indian business" by unilaterally abrogating treaties and abolishing reservations. House Concurrent Resolution 108 in 1953 was a major step in this direction, and during this phase of the project 109 reservations were slated for termination. Concurrent legislation was formed to encourage families to move out of their poverty-stricken land bases. Communities fought this in the courts, with most succeeding, but this period created a massive Native diaspora, with a minimum of 100,000 being removed to major cities. While some returned, critics argue that this period was instrumental in forming today's reality that two-thirds of Native Americans live in cities.

"No one ever thinks about it, but the question is, 'Why would we leave the reservation?'" Shannon went on to answer the question. "Because it was so poor. The environments were so poor. The water system. There was no infrastructure. The houses we had were dilapidated. I remember growing up, even when I was really little, in a mud house with rooms that were divided by cardboard."

Shannon, a member of the Akimel O'otham (River People), talked with us from his humble office at the Indian Health Center in San Jose, California, in October 2020 as fall set in across the continent. Born and raised on the Gila River Indian Community in southern Arizona, Shannon grew up in the life of financial poverty that so typically besets many reservations in the United States.

But since then, he has gone on to become both a delegate and participant at the United Nations Permanent Forum on Indigenous Issues held at the UN headquarters in New York City. He has been a co-chair for the Global Indigenous Peoples Caucus, and has conducted and hosted lectures on the UN Declaration of Rights of Indigenous Peoples (UNDRIP) at the State Capitol in Arizona and for Arizona State University, along with doing so for numerous local colleges in Arizona and California. Regularly active in political and social justice actions in Central and South America, Shannon was also instrumental in assisting the actual implementation of the UNDRIP. For quite some time now, Shannon's focus has been on his work as a Native American Traditional Cultural Advisor to the Indigenous inmate populations at the county, state, and federal prisons in Arizona.

Despite the extreme poverty of reservation life during his youth, Shannon's parents decided to stay. Meanwhile, like so many Native families, many of his relatives left and spread themselves across the continent. His aunt ended up in Daly City, California, cousins moved to Los Angeles, and a great-uncle ended up in Chicago.

"People were moving all over the country, people were leaving. And the reason why that story is significant, is that even *then* environmental and climate change were impacting those of us living along the Gila River. When the European settlers came, they started damming our rivers upstream, in towns like Florence, Arizona, and even further northeast of us."

But even before that, policies from Presidents Roosevelt and Coolidge created dams along the Gila River, which started diverting the water. "My people were dependent on that water, and the damming happened seventy, eighty years before my parents were even born," he said.

As a direct result of this, his great-great-grandparents remembered when whooping cough wracked their community. "They called it a dry cough, and back then they didn't know really what it was, but there was development in Phoenix and Tucson, and destruction of those traditional ecosystems, those plant systems that would keep the soil in place and would keep the dust down. And this dust later became the cause of Valley fever. People will say, 'Well, what is Valley fever?' Nobody really knew what it was, and we know that it's a respiratory issue, but we believed that it was connected to the destruction and degradation of those soil systems and the traditional plants." In this way, settler colonial policies and an incipient capitalism had already begun harming and displacing Indigenous people in the southwestern United States.

As Shannon grew, he watched his parents working, but said, "They weren't really making money." His father was earning roughly $150 per week, and this was supposed to provide for nine people in their home. "We were all Native kids, poor, and we would get our clothes at the local church nearby. So for us, it's not just about the environment. It's the cultural and spiritual integrity. But for me as a young kid I didn't think about the environment. We had culture, but as I got older I wondered how my parents held onto that."

Many of his people were no longer growing their traditional squash, beans, corn, and melons. Ceremonies were dying out, which only more recently have begun to be brought back. "So for my family who relocated, they went for a reason. They went because the water wasn't there, there was no more livelihood, and the government was touting it as, 'You can relocate and you can have a great life.'"

It was a time when most of the Natives who were moving off of the reservations and into the big cities were finding themselves in poverty-stricken areas, and other places they did not want to live, in addition to suffering high rates of alcoholism and drug addiction. "There was a lot of loss of culture at the time. I always wondered why my parents stayed in such a dry place." Born in 1966, he remembers there being "nothing" there, while the White man was engaged in constant development of farms along the Gila River.

"We weren't growing food for ourselves. The tribe was growing alfalfa and hay. Rarely were we growing anything that was for our benefit. It was mostly for economic reasons. We weren't looking at our own community." Instead, his family was getting by on eating government rations. Shannon laughed while telling us he grew up on this kind of fare. "I think my favorite was the canned meats. I don't know if it was meat, but it was pretty good." We all laughed as he described the long cylindrical cans of pineapple juice that taste half like metal and half like juice. "And of course everybody knows about the huge blocks of cheese," he added. "You could throw that on the fire and they'd never melt."

Shannon came back around to the serious question of why his parents, despite the hardships inherent with living on the reservation, didn't leave. "I asked my mom, and it was because of trauma," he explained, his voice deeper and softer, "as a young woman. And, it'd be remiss of me not to speak about the church and the impact of the church within our communities with regards to the environment. You gotta remember when the church came in, when the Christians and the Catholics came in, we were told not to do those things that were traditional. We were told, 'If you just pray, and if you live like the White man, you're going to be okay.'"

At that time, President Roosevelt was attempting to move all of the Akimel O'otham people to Oklahoma, but Lion Shield, one of their last chiefs, refused to move, choosing to remain in the desert. "His idea was about human rights," Shannon said of the chief's decision, before sharing the core of why the chief made the decision he did. "Don't you see me as a human being?

"And they said, 'No, we'll give you this land then, but we're still gonna take your water.' And since then, for roughly a hundred years now, the Gila River has been dry. So what happens is that the water table dropped and people started cutting all the mesquite trees, cutting them down to sell for wood so that they could just survive." Shannon cites a *Chicago Tribune* story about Indians dying from starvation in the Arizona desert.

"Thousands of us were dying; four thousand had died in 1910," he said of the story he found while doing research for his thesis. "That was from environmental destruction of not only my community, but across the Southwest." He paused, then came back again to the question of why his parents, despite all of that, chose to remain on the reservation.

"My mom said that she would not leave because her trauma from dealing with Christian and Catholic people was so devastating that she was afraid of them," Shannon explained, his voice quieter still. "She was afraid. But that's my mom's story, and I'll let her tell that, but there are reasons for our young women being afraid of these Catholic and Christian men who were out there at the time. And my mom was a little girl when this happened to her, so my mom never left the reservation. Never. And she's seventy-plus years old now, and she'll more than likely die there. They tried to take her to a Christian school and she didn't have a good experience."

When Shannon asked his stepfather why he went to Korea, he said, "I went because there was nothing here. There was nothing here. You couldn't work. We drank all the time." Shannon paused, took a deep breath, then continued. "That was kind of the era of them drinking. There was this blanket of despair that was across the reservation, and around the community. One old man told me many long years ago when I was pretty young, 'You know what the problem is with us, nephew?' and I said, 'What?' He said, 'We are using a blanket that the White man gave us. And that blanket we assume is comfort. And that blanket has holes in it. And no matter how we try to pull it over us, it has holes in it and we're still cold. We can never warm up. We can never heal ourselves. And we keep pulling that blanket, and we keep fighting for the same blanket, and that blanket is despair and sadness.'"

Shannon returned to the question of why his stepfather went to Korea. "He said, 'If I was going to die over there, your mom was going to get my military pay,'" Of his stepfather's response, Shannon said, "I think it was

150 bucks or something like that, something silly at the time. But that was a lot of money for us back then."

His family existed on government commodity foods, augmented by beans, tortillas, and hot dogs when they were lucky, and this stands as the example Shannon used to show what happens to Native people when their environment is destroyed or altered. "There is a legacy of environmental destruction that has happened since the settlers landed in 1620. Since they came from the south in the 1500s, there has been a legacy of environmental destruction," he said sternly.

He spoke specifically of the destruction of their waterways other than dams, describing how when the "Puritans got off the boat," the rodents in the boats escaped, entered the river systems, died, and poisoned the waters, then it poisoned the people. Similar phenomena occurred when other settlers and colonists from various countries brought chickens, pigs, and cattle, which brought diseases as well as destroyed water systems and land in various regions.

This brings the destructive impacts upon Indigenous People from climate disruption into a new context. Across the continent, and across the world, they have been having their lands destroyed for generations, suffering varieties of environmental collapse, wrought by settlers and colonists, and only recently, by the broader climate crisis the settler-colonist mentality set in motion when it began.

What Shannon shared about his family is reflective of what is happening in many reservations across the continent. The majority of Native people are leaving because, as he'd mentioned, there is nothing there for them. This is exacerbated by the fact that those systems are antiquated. This fact was underscored by how Indian Health Services was ill-equipped to deal with the pandemic, and how many of their education systems still teach the Western ideology of conquest, of Columbus Day.

Yet the young people are also leaving because of the increasing environmental destruction. "If you go up north on the Navajo Reservation, the water table for the folks in Big Mountain still remains, but we know that there's hardly new water anymore. Climate there has changed. The San Juan River is pretty much the only one not dammed, but they're looking at damming that up too. They're destroying everything in that region." Shannon considers the Gila River as his home, his place, but he too left because "there was nothing for me there. There was nothing there environmentally, and there were very few cultural ceremonies." Most of his friends drank, and crack cocaine was coming into their communities, along with heroin, which had already been there. "People were trying to find themselves, but not knowing what they were looking for because it's all been so displaced. I'm one of the lucky ones who got out."

But his tribe has been fighting back, and they are one of the first to fight for their water rights. In 2008, President Bush signed an executive order returning their water to them, and since then the tribe is now asking community members to develop their own farms with it. Many of the younger people who are leaving are getting more education and finding better jobs, but despite this, disparities among Native communities on the reservations remain devastating.

Given the rupture and disconnection he shared with us, we asked Shannon if he would talk about the healing work he is engaged in. During the 1970s, his teachers, whom he refers to as his uncles as a sign of respect, people like John Funmaker, Archie Fire Lame Deer, Crow Dog, and others, began holding sweat lodge ceremonies inside prisons. San Quentin Prison in California was one of the first places to hold a sweat. Shortly after this, the American Indian Religious Freedom Act was passed, and Natives in prisons were allowed to participate in sweat lodges. Roughly a quarter century later, Shannon was asked by his uncles to consider going into the prisons to run sweats. Initially, he found coordinating sweats in maximum-security prisons to be challenging, but has now been doing it for quite some time. Taking traditional sweat lodge ceremony into the prison systems of Arizona and California, Shannon described the work as having been "beautiful and amazing." He went on to share a story of a man he met in prison who recently passed away.

"He went into prison when he was twenty, and by the time I met him he was seventy. He was a Lakota man who had never sweat before and wanted to come into the ceremony. He was old and had a lot of health issues, but he came in. He had killed some people. He came in to seek forgiveness. I told him, 'Don't ask for forgiveness from me because I can't give you that. I can *personally* forgive you. I can pray with you.' But his healing was so transformative that soon after that he passed away. He felt that he could finally move on."

Shannon paused, then continued. "So the work within the prisons is something that's profound to me. I go in and I see these men who have never learned how to pray. Some of them know a few traditional songs. But it's very rare that you find Native men who have stuck with ceremony and are committed to it."

Shannon believes healing comes from the community and that that level of traditional and spiritual teaching needs to be maintained. "It's our communities that keep us out of prisons. And I always tell the men, 'Long before you were in prison, you were in prison behind these other walls. You were in prison on the reservation. You were in prison long before you committed a crime. You violated our spiritual law, and you violated our traditional law. You should know what our traditional laws are.' And some of them don't. So you have a lot of these young men and women in prison who have been violating the natural law long before they entered into these systems of incarceration. They have devastated their families, they've devastated their communities, and they devastate the lives of their family and children."

Shannon talks with them about "natural law." He talks with them about the balance of natural law, "because trying to be harmonious is the most difficult thing. The balance is already there. It's us as human beings that are destroying this balance. It is us, the two-legged, that are constantly greedy and looking for something that can move us up the scale of economic stability." But he explained that this is not one of our natural laws. We cannot overconsume and expect things to grow back. He references how many of the men currently incarcerated for murder were drunk, high, or both when they committed the crime.

"In 1799 a guy by the name of Medicine Lake talked about 'mind chang-ers.' He talked about alcohol being a spirit. He said that the spirit would come in a liquid and it would destroy our communities. For centuries al-cohol has laid waste. He said, 'It'll create a pile of bones.' And it created a pile of bones, and those bones are our people." Shannon included drugs with this prophecy, and asked, "What does it mean to heal when you violated our traditional laws? How do you heal from that?"

What he then discussed is not of the Western concept of forgiveness. "It's a concept of balance for us as Native people. How do we balance our-selves? Because the concept of forgiveness is the White man's way of ask-ing God for forgiveness. It's kind of silly for me to ask God for forgiveness when God, our Creator, is forgiving us every day for atrocities that we're doing. So we as Native people need to look beyond the concepts of a White, Christian, male patriarchy, because that destroys our balance with Mother Nature. Natural law influences our thinking, and this is why we must go back to it."

Shannon explained how his ancestors used to tell his people to be in bal-ance, and then harmony will come. But you must first be in balance. "If you don't know how to be in balance, learn the songs, learn the ceremo-nies," he said of what he works to bring into the prisons. "When you walk into the prison, you see these guys all out of balance. There's no harmony in the maximum security prisons, there's just 'Watch your back.'"

Shannon spoke to the tension that exists between certain Native groups, between Whites and Blacks, and between those northern and southern Mexicans who aren't aware of the fact they are all Indigenous peoples who have unconsciously destroyed their natural balance. He described the prisoners' ongoing struggle for power and economic gain as akin to everyone being in the sweat lodge and wanting to get out at the same time because it's so hot.

"They're trying to crawl out instead of just humbling themselves and bowing their head and saying, 'You know, I can do this because even though it's difficult right now, the right thing to do is to pray and to suffer a little bit and to give of myself, because I have taken too much. And in that, I have to withstand this heat.' And I'm just talking metaphorically, but their idea of balance is so out of whack that you go into the prison and they fight about who's gonna run sweat, or if the gay and transgender folk get to use the wood. Yes, they do, they get to use the wood too! See, this kind of Christian-dominated understanding of separation has destroyed our community, and it seeps into everything: our political systems, our cultural and traditional systems, and even into our traditional ecological understanding of how we relate to Mother Earth.

"So we fight about the small things. We fight about who's gay or who's transgender. Personally, as a spiritual leader, I don't care what you are. I care that you're there to pray. Christianity has seeped so deep into this hatred toward one another that we use that colonial concept of what's right or what's wrong, and we let it seep into our cultures and tradition. We have gay and transgender people in our community, of course. The idea that somehow we could hate our own people, that is the 'out of balance' I'm talking about."

Shannon cited the infighting within the prison systems as another example of how out of balance people are, but then uses other examples that implicate most of humanity. "We're so out of balance that we will negotiate around whether it's okay to cut through one of our sacred mountains back home to build a freeway. We're so out of balance that we allow the government/state actors to come in and destroy our rain forest and our water systems." Nevertheless, Shannon pointed out that there remain people who continue to fight against that. "There are people that are saying, 'No, you are not going to build a pipeline here. No, you're not going to build a wall here because we have a relationship to the land and to the animals and to the plants. You cannot do that because what you have done for centuries has destroyed our balance. And we have to maintain that balance.' The four hundred million Indigenous people around the world that are holding onto those ecosystems, they are the ones that I believe have that moral compass."

Shannon continued to talk about the critical necessity of balance. "This pandemic has shown how weak we are as a country, how weak we are as people, how unbalanced we are as people that we would argue against something like a pandemic and say, 'No, I don't want to wear a mask because it violates my sovereignty.' No. If you had balance, you would understand."

He cited Indigenous people as examples of balance, and again came back to the importance of helping Natives in prison find their inner balance. "Indigenous people know. On the reservations, when I go home, everybody's wearing a mask. Everybody's careful. They want to protect the elders because the elders are our libraries. They carry our stories. They are the song keepers. They are our teachers. They are the ones with the PhDs and they don't come from UCLA or Harvard. They come from the Earth, and they carry that PhD of understanding what it means to go pick by a certain time, the idea of what it means to go hunt during a certain time of the season. The idea of what it means to go fish in this area, but not that area, because of its lack of abundance. The idea of why the water is important to flow in this region. So, yeah, I love that work. I think it's so needed. It's so amazing to me that they're not there just to get out of the cell when they come to the sweat lodge. They're there to pray."

Stan thanked Shannon for all that he'd shared and for his ongoing work in the prisons. He asked if Shannon would speak about the Western world's grasping toward so-called Traditional Ecological Knowledge, now that Western science has finally understood the ramifications of Western life being so grossly out of balance with Earth.

Speaking from his experience in obtaining a degree in environmental science, Shannon recalled studying the change in soil samples taken from undisturbed areas compared to areas where reclamation projects occurred. Specifically, he did this on the mountain where the Snowbowl Ski Resort is located in Arizona, and found that reclaimed water from developed areas destroyed certain plant systems on the mountain. The owners of the resort wanted to expand the ski area, but the Navajo, along with thirteen other tribal nations, believe the mountain is a sacred place. They believe the Kachinas (spirit people) go to the mountain to conduct ceremony.

Shannon's research revealed, specifically, how the microbiomes of the soil on the mountain were impacted for the worse. This is due to the fact that when water from an unnatural source that does not come from snow that falls on the mountain was introduced, it changed the entire system. "This confused some of the spirits," Shannon said. "Because we're interfering with the natural system of Mother Nature, and the story of Creator in that contact between the Spirit and the Kachina Spirits, and the Spirit of the creek." It is yet another example of humans impacting the natural flow of things.

For the last five years Shannon has been working with and supporting the Amah Mutsun Tribal Band on California's coast in the Santa Cruz area. The tribe has been maintaining the forested areas of their region by way of fire control measures, and regrowing traditional plants that used to grow there. Not long before we spoke with Shannon, wildfires had devastated much of that region. A week before the fires came through the areas stewarded by the Amah Mutsun, Shannon asked some of them to clear a particular location. Brush and trees were cleared, and they built a sweat lodge. "Then the fires came, and they destroyed the buildings, and the sweat lodge still stood, along with the trees that were intentionally left around it. I'm not patting myself on the back. I'm using traditional methods, ecological knowledge that Indigenous peoples have used for millennia that maintained their ecosystems and allowed their ecosystems to be abundant every year, to grow and to resurface again, whether a fire came through or not, whether it was a lightning fire or man-made fire."

He stressed that it is time to listen to the people who know, and pointed back to the mountain in Arizona. "Listen to the Hopi, listen to the Navajo who tell us we are destroying the natural balance, the natural law of our spiritual relationship to that mountain. As a human being, you're trying to interfere with something that you shouldn't interfere with, just to expand a ski resort, just to increase your economic growth, just to increase economic development in the city of Flagstaff."

Shannon pointed to the contrast between how non-Native people see land as a resource linked to the economy, while Native people see it as a means of survival, and a way of living "for," rather than "against." "The United States is roughly 5 percent of the global population, yet we consume at least a quarter of the planet's goods. We are destroying ecosystems around the world, and we're putting in governments and dictatorships that say, 'Oh yeah, we agree with you, keep cutting those trees because we need our nice furniture. Let's create this massive farm in the Amazon and cut down all these trees and build these cattle ranches.'"

Shannon contrasted that with how Native people used to hunt buffalo and other wild game in a way where they took only what was needed. "That traditional ecological knowledge is not just based on ecosystems, it's the way we try to balance ourselves with the animals and the plants and the water systems upon which our livelihood depends. It is how we survive, and how we *have* survived."

Shannon lamented still another form of how out of balance most of his people have become, by way of poor diet leading to diabetes, hypertension, or high blood pressure, coupled with a lack of healers, as less of the old ways are being passed on to younger generations. He wondered if it is possible to get it all back to where it was.

Thinking about this for a moment, he then said, "I would say yes, but it's gonna take some work from all of us."

I sat in silence, taking it all in, his words and information heavy on my soul. He asked if I was still there, because I had been sitting so still, hanging on every word he shared. I tried to explain this to him, and he laughed and told me he understood.

We asked Shannon if he would speak to how much despair there is among so many people nowadays, and more specifically, how we should comport ourselves during these times. He spoke of going to Navajo Nation, to Big Mountain in the late 1980s when Hopi and Navajo elders were fighting against the Peabody Mine after the federal government partitioned the reservation to help the mining project move forward. The Hopi and Navajo had been fighting the issue long before that, but from that point on, he watched the federal government's implementation of "divide and conquer" with the Hopi and Navajo. "Can you imagine two nations that have lived next to each other for centuries," he asked, "now purportedly fighting against each other? 'This is my land, this is your land.' Again, this Western concept of ownership, right?" He lamented the effectiveness of the government's divide and conquer strategy, given the two Native nations used to annually share corn, beans, and other necessities.

Shannon reiterates the concept of ownership as running counter to the natural balance of who we are, and what we are. While he is inspired by the awareness and energy around the current social movements and the Indigenous demands for social justice, he asked, "For them, what does justice look like? It's still justice in a colonial system.

"Now, no matter who we vote for, Biden or Trump, Indians are not going to see any benefit. We're still just waiting for the crumbs to fall, because we're not at the table. Maybe with a Democrat they'll give you something, but with the Republicans, people are just fighting for the rights of just being human." He pointed to the fact that Native people weren't even allowed to vote until 1924, because they weren't seen as citizens in their own country. "So this idea, then, that these young people are pushing their agenda is amazing, and it's great, but they need guidance. And that guidance is so you don't fall into a colonial system that has not been working. We need to revamp the whole system."

Shannon mentioned numerous other political parties and challenged what has become a two-party system. He believes every party should be involved, and settler colonialism must be discussed, reminding everyone of its violence, not just toward poor people of all races, but Native people, as well as immigrants coming across the borders looking for economic opportunity.

Shannon is proud of the young people "pushing the street," as he put it, and laughed at those who are calling for them to be arrested due to the sporadic looting that has been happening. "We've been looted for 528 years. This looting has been going on . . . we need to burn that shit down." Shannon paused, then clarified that he meant that metaphorically. "Burn it down to rebuild it again. You have to rebuild."

Shannon cited the need for building different things, like community centers and gardens, which are actually going to take care of people and their families for generations to come. When he spoke of justice, he did not refer to the U.S. Constitution, but to a justice born from the Iroquois (Haudenosaunee, "People of the Longhouse") Confederacy of upper New York State and southeastern Canada. The confederacy is known as one of the oldest participatory democracies on Earth.

Shannon cited the Two Row Wampum Belt Treaty, formed by the Iroquois Confederacy with the Dutch government, that became the basis of all their subsequent treaties. "That's the treaty where the Indigenous people said, 'Okay, you're on that side of river, we're on this side, and we're going down the same river. You're in your boat and I'm in my boat. We can go down the same river. But if you try to cross my path, you are forcing your ideology on me, and you tip over my canoe.'

"They have been tipping over our canoes and they've been walking on our path. They've been trying to stomp our path and burn our path down. And so our idea of justice doesn't come from the Western concepts of justice. Our idea of justice comes from a traditional ecological balanced knowledge that comes from ceremony; it comes from songs, it comes from the teachings of my mother and my father and their grandparents, who didn't speak a lick of English, who spoke only O'otham."

Shannon believes all of this is missing today. He sees how social justice movements led by younger generations are having an impact, but offered a critique. "It's not just about George Floyd and Breonna Taylor and all these other young folks who've sadly been killed, but their [protesters'] idea of justice is still White justice. Taking them to court, taking them to jail, I see that, but you're just going to erase them [the perpetrators] from the community, and that doesn't bring the healing of Breonna Taylor and George Floyd's families. What does it bring? Sure, it's going to bring some sort of a sense of justice and healing. But like for my mother who lost many children, that takes a level of prayer, a level of community, a level of relationship building."

Shannon doesn't know if this can be accomplished today, so he took us back to the analogy of the Two Row Wampum Belt Treaty of the Iroquois Confederacy. "They've been creeping onto my side of the river and bumping my boat for over five hundred years. They've been pushing me over, and telling me I can't go downstream anymore. They left us behind. So, as Indigenous people, we are left fighting for scraps: scraps of justice, scraps of food, scraps of a seat at the table, scraps of everything else that impacts our lives."

At that point he paused for a moment of thought, then continued. "But the thing that makes us strong, I believe, is that we have the moral compass built in. I believe that traditionally we have this compass, and there are people around the world who have it today. Evo Morales, while we may disagree with some of his politics, is really pushing the idea of what it looks like for Native people to survive today. He believes we can balance our ecosystems with our economy. I believe there are Native people out there to this day, whether they are at Standing Rock, or on the islands of Hawai'i, who are trying to maintain that balance.

"And they're up against this huge machine of devastation, and that devastation is capitalism. Capitalism is a machine. It's hungry and we keep feeding it, and we're starving. We're starving for something else other than capitalism. We're starving to survive, and we're starving for that balance. We're starving for that prayer. We're starving for that song."

Shannon was reticent to continue that thread, saying he worried about sounding cynical, but he spoke of the need to face the hard realities, alluding to dire endings if capitalism continues to be fed and to dominate the planet as has been happening. But he cited those, including himself, who have been fighting, and continue to fight. "A warrior is somebody who fights, and somebody that knows what to do," he said firmly. "There's a difference between a warrior and a soldier. A soldier follows orders. A warrior knows what to do. We are ecological warriors. Mother Earth will balance herself. Whether or not we get to see that balance, and whether or not it will happen in our lifetime, we don't know. In the meantime, we have to slow the growth of this capitalist machine, and we have to stop it. We have to say, 'Wait a minute, we're destroying what we're trying to build for our children.'"

Shannon has nephews and nieces that he loves deeply. He wants them to see something different from how the world is today. "This probably won't happen in my lifetime. Justice is a slow process, right? Justice through the lens of Indigenous peoples that have that moral compass, that have that idea of balance, that have that idea of relationship building and interconnectedness between the Earth and our love for our fellow two-legged and four-legged and winged brothers and sisters. That's the moral compass that I believe that we bring into the prison systems, and that we bring to the reservations. I believe that we can bring it to people around the world."

We all sat with that, quietly. Knowing how busy Shannon was, we thanked him for his time, before Stan asked him if he had anything else he wished to add.

"I started out talking about my parents. They got sober around the time I was twelve. Now, when you get sober in Indian country, it's not like you got AA around the corner, right? So you still have a lot of issues, and you still have a lot of dysfunction. My family was always struggling. So when I was fifteen, I was sent to a Christian school and they cut my hair. They said that in order for me to look presentable to God, I had to wear collared shirts, and my hair couldn't go past my shirt collar. That was their idea of being presentable."

Shannon remembered when he came home from that school for the first time. He hadn't seen his uncle since his hair had been cut. "He asked me what happened to my hair, and he started crying. This was one of my uncles who was involved with the American Indian Movement (AIM) and the Red Power Movement." Shannon told him why the school had cut his hair, and his uncle, as did his mother, became quite emotional. "My uncle, who had long beautiful black hair, and was a tall, handsome guy, said, 'We'll never assimilate that way.' He said there were warriors that have lived throughout the centuries since the settlers came, and however they wanted to wear their hair, that's how they wore their hair."

Shannon had a little fear of his uncle, who was a powerful person in AIM, and was a security guard in the movement. "He was always saying,

'Fuck the White man.' He was a hard guy, and he was crying about my hair because my hair was cut, not by my family, but by some White men at the church at this Christian school, because I had to look presentable to God." Shannon's grandmother was still alive at that time, and told him he was beautiful to them, no matter what. She did this to try to make all of them feel better.

"But just through that idea that I was forced to have my hair cut, the force of that memory of what was done to all of us was brought back to them. This trauma was brought back to them, right through me, through my hair. But no matter what, my family maintained this level of balance and tradition, despite all the traumas and the struggles that they went through, from my dad, working in the field, and my mom not wanting to leave the rez because of her fear of other outsiders."

But his family maintained their dignity, despite all of that. "They fed me, and fed me beans, even if it was four or five times a week. They survived. They maintained that level of balance, what they knew to be balance. And so now I have my hair long." Shannon leaned forward in his chair and pulled his long ponytail around to his chest to show us.

"I walk around with it, proud all the time. People ask, 'Are you Native American?'" he said. After a pause, he smiled and slowly leaned back in his chair. "I'm not Native American. I'm not American Indian. I'm older than that concept. I'm older than that concept." He repeated and paused, before making his last comment, very softly, with reverence.

"My family did what they could, and that's why I sit here today."

11

Edgar Ibarra (Chicano, *Yoeme, Tarahumara*)

Healing

COMPOSED BY STAN RUSHWORTH

These traditional ways are what's going to help us reclaim our neighborhoods.

—Edgar Ibarra

Edgar Ibarra (Chicano, Yoeme, Tarahumara) is a student at the University of California, Davis, and an active member of the Milpa Collective, a group working to end mass incarceration, dismantle the school to prison pipeline, and build community and people power through introducing Indigenous values and practices into communities. He is a good-looking young man, very fit, with a ready smile and a laugh that erupts quickly after a very serious train of thought, or that precedes his softer reflections and speaking. His words are genuine, real, each and every one, born out of years of contemplation.

I begin by telling Edgar our approach to the interview: "This is like the 'Palabras' you do in your community circles. You have total freedom and total say. Our focus is the disruption of Earth; how did we get here? How do we move on in the right way? What's in the way of that? And no matter the outcome, how do we carry ourselves? Where do these questions speak to you? You have the eagle feather, so it's your call."

"Okay. I appreciate this way a lot, so to start I'd like to introduce myself. The language is becoming more and more important for me, so I would like to start off with that, trying to reacclimate myself with that language. It might not be Yoeme or Tarahumara, my ancestors' languages, but it's close, something that might've been understood during trade at one point or another.

"Piyalli Cualli Tlanezi na no tica Edgar Ernesto Ibarra Gutierrez. [Hello. Good morning. My name is Edgar Ernesto Ibarra Gutierrez.]

"I also want to acknowledge my family and where they come from. My mom's name is Alma Rosa Gutierrez Galvan. Due to the revolutionary war in Mexico, her family has roots in a lot of different places from northern Mexico, but they settled in Mexicali, where my grandfather worked on the train tracks connecting Mexicali and Sonora back in the forties. My dad's family come from Metzquititlán, now called San Juan de los Lagos in the state of Jalisco. I acknowledge the Indigenous component of who I am, but also the European component."

Edgar reminds me of the tremendous diaspora of Indigenous all across the Americas, a key view that many forget in holding to newer borders without thinking of how recent they are, and how they've affected so many older peoples of the continent.

"I want to bring in all those things because it can get crazy trying to figure out who you are, how you can reject one part of you, which is rejecting yourself, and I'm trying to come to terms with who I am in my totality."

Edgar then enters into his experience of going to college right now, and what it means for him to be around people who hear him. He emphasizes that there is no separation between the issues of climate change, colonization, and his life. His life is an embodiment of climate change and the gamut of desecration surrounding it.

"Again, I appreciate this opportunity. I'd always been told, growing up, that English was not for us. That it was for someone else, so having a teacher come through and tell me, 'You have a good way of writing,' opened something for me. I was twenty-five at the time."

He pauses, collecting it all together, smiles, then nods, engaging us with his eyes. Behind him on the white wall of his small apartment, a bright blue painting of a golden eagle, perched upon a stone carving of Quetzal-coatl, shines among other Indigenous shapes and symbols of Mexica.

"I'm going to give a little bit of history of myself and then lean into the questions as far as they take me. I grew up in Watsonville, but I was born in La Mesa, California. My mother had me over here at her brother's house, and then she went back to Mexico. My father told her, 'You got to go have him over there, so he can have a better opportunity.' I didn't know this story until last year. So she came up here by herself and left three young ones back in Mexico. She had me and made her way back to Mexico." I think of the ongoing diaspora again, and the reasons for today's migrations.

"A year or two later they came back up to the States and my father was incarcerated. He ended up spending thirteen years of his life incarcerated, most of my life, you know, which in turn impacts me because I also ended up spending about ten years of my life behind the walls, from fourteen to twenty-four."

He is quick to add with a smile, "But growing up in Watsonville wasn't 'behind the walls.' It was almost like a shock sanctuary. The Chicano pop-ulation has called it the Aztlan of California, where everything was hap-pening. Danza would happen in the plazita, where they would have the rites of passage for young girls, with hundreds of dancers out there, and this was in the eighties. Then they moved it to Pinto Lake Park, with hundreds of people coming to witness the rites of passage and coming of age of a young lady." A broad smile of appreciation crosses his face, punctuating the importance of the whole community honoring the young women. He gestures with his eyes, making sure we know this, that we get this.

"These images are in my head," he pauses and goes into detail, "and back then I kept thinking, 'It's the Indians' house, Indians and Indians.' I walked through Watson' and witnessed the murals that adorned the schools in the city, and I thought, 'What's going on? Why do they have the homies right there next to these Indians?'" He shakes his head, laughing and incredulous that he didn't put it all together as a child. Then he goes into the other forces at play that prevented the connections from bearing fruit.

"At the same time, growing up in poverty, I witnessed violence, systemic violence, prostitution, drugs, everything; you know, we witnessed all that. I watched uncles deteriorate to heroin abuse and addiction, and get lost in that whole mind-set. I saw single mothers with a lot of the husbands nowhere to be found, raising anywhere between three to five children on their own, you know?

"Sometimes those children who were fortunate enough to even have fathers, still had an absent father, because they were just 'gone,' strung out on drugs or alcohol. This is what we saw in our communities, even downtown. But living closer to the levee is where you saw the impacts of poverty much more, people living clustered in a one- or two-bedroom home with ten to twelve people.

"You saw the impacts of climate change on the levee too, like today, the flooding that took place, and the amount of trash all around, because a lot of people resorted to the levee as a safe haven to go to, so if you were homeless, that was the place to go and sleep."

Those facts of childhood stream out of Edgar, and he puts them out as just that, straight talking, then the smile comes again.

"But there was also some sort of security and safety growing up there. We would make our trek as a bunch of kids from the neighborhood down to the river. And underneath the bridge where the train passes, there was this big old swing where we would jump into the river, and we never noticed why there were shopping carts.

"We never sat there and asked, 'Why is there so much trash?' We just played. We played, and we didn't even take into consideration the runoff from the agricultural businesses, with the strawberry fields and everything right there. We just played in it.

"When we were growing up, we never sat and considered our situation. Sometimes I've been asked, 'How was it growing up poor?' To be honest, I was just going around playing with my friends, playing marbles, you know, playing childhood games out there."

This back and forth is what he does in his reflections, weighing one side of life next to the other, the hard side next to the soft side. He remembers the good of childhood amidst the bad, then goes back to the trajectory of where it runs for too many people. He played and saw the good as a kid, not the bad. But then it shifted.

"I think it was the first time that we got pulled over.

"I was twelve years old. They sat us down and made us take off our shirts to tattoo-check all of us. We were twelve years old, most of us, in between eleven and thirteen. And at the time we didn't understand, but in retrospect, we were being profiled for who we were and where we lived."

Edgar goes into how everything began to shape into something very different then, very quickly. He describes the thinking that developed in the children.

"A lot of these things have to do with the mind-set we grew up with after that. We didn't really have a sense of self-worth. We thought our life was short. We believed that it was meaningless. There was really no clear purpose. And what we saw was that we had very few avenues, right? It was that you were going to work in the agribusiness as a laborer, work in the canneries, end up in prison, or pass away at an early age. That's what we saw."

He shows how the whole situation, the economics, the vision of him and his playmates by a surrounding culture and economy, worked to develop a philosophy in the children early on, a point he sums up strongly.

"We had a fatalistic idea of everything!" He stands hard on these words, eyebrows raised, and I think of this in children, remembering how many young faces I've seen carrying this fatalism here and in Central America.

Then he pauses to reflect again, quietly. "In retrospect, where I am today is pretty amazing. I was told by the vice principal of my middle school that I would be in prison at age eighteen or that I would be dead by then. My mother didn't understand what the woman said, so I had to translate that for her, translate those words. It's one of those things you think about in your life, and I held that against my mother, because she didn't even know how to advocate for me.

"I'm still young, only twenty-eight years old, thinking about so many things I held against my mother, but then I come to things like, 'She didn't know English. She didn't know what she was getting herself into.' And neither did I. She was by herself, undocumented, scared, thinking about what was going to happen if she got deported. What would happen to her children? These are the things that happen growing up in this type of environment."

Edgar pauses again, flashes a smile of obvious gratitude, shaking his head, then goes to the other side of it once again, always both sides, seeing everything he can, balancing it all.

"At the same time, there were so many blessings, so many blessings. We had the love of our grandmothers, of the seniors always looking out for us, always making sure that we ate, always telling on a bad behavior. 'I'll tell your mom. I know when your mom's coming home.'" Edgar laughs at that memory, then quickly sobers as he continues.

"But with a lot of my friends, the traumas hit them hard. And the only way they could cope with them was with alcohol and drugs, and that took them for a crazy spin. You couldn't even recognize them anymore. They were lost. Again, it's the sense of not valuing yourself. Not being able to say 'Okay, this is who I am, this is the potential of what I can be.'

"I can only think of how few people there were in our lives at that age that really took the time to tell us, 'You matter.' Sometimes our parents didn't even know how to articulate that. Again, I share this because seeing all of this together makes sense for me in my own moving forward."

A silence comes between us all, and we wait, nodding at each other across the Skype distance, and the distance melts away. I think of how what he is seeing and saying is a key to knowing how climate destruction and social injustice feed each other. Like Edgar views his own life, we have to see it all to move forward, not just what we want to see. Edgar leans forward and looks at us with calm intensity, relating facts, words rising out of heartfelt contemplation, out of experience.

"At the age of fourteen, I would commit a horrible crime that would impact my family and my victims. I was incarcerated for the next nine years, eight months, and four days, navigating the whole juvenile justice system, aka 'Gladiator School,' then to the county jail, then into the prison system.

"I hit one of the worst prisons known in California, right in our area, Salinas Valley State Prison. At the age of eighteen, I was on a level four yard, one of the worst prison yards you can ever be in. The first sign you see is 'No Warning Shots.' If anything happens, it's like you're going to get killed one way or another, and at the time it was the most violent prison in California. I was eighteen years old going into one of those places, turning nineteen in there."

He shakes his head, a young man having spent a big piece of childhood behind those walls, inside that life. More reflections pour out of him, calmly and passionately, uncomplaining but naming it.

"Going through that system was dehumanizing. It stripped us of our dignity." He has reflected a lot on his life, but he reflects now on the present, on right now. "I'd never thought about some of these things, but the quarantine has allowed me to reflect, to dig in deep and look at how I'm going to move forward. What's the destination? Where's the goal? And I'm having to realize certain things that I was not comfortable in expressing even to myself, and none of it to others, like that I spent nine years, eight months, and four days being abused by the system, having to strip down in front of people, having to shut down every single thing in my mind and body, just to rationalize that this was okay.

"I shut down a whole big part of me, and I didn't realize that until lately, through having conversations with other folks that have gone through it, having to address that, as young men, we were treated not as men. It was sexual abuse; to consistently be degraded in front of hundreds of individuals on the yard takes something from you."

He gathers it together again, how we see and treat people within the larger "civilization." It's about desecration. "So prison put something into perspective for me. 'This is where we throw away our trash.' You know, 'This is what we think of you. We're going to throw you away here forever.'" It's another step of the trash on the levee, about disrespect of oneself and others, a self-perpetuating thing, and a logical continuation.

"Everything that was fed to us there was bad, literally chaotic, being in a closed space with all these pent up emotions up from different people who don't know how to navigate them, and you feel it, and it's crazy.

"I share this because it's important to where I am today, being able to overcome that and go beyond it." He shakes his head and takes this idea to another level of understanding, quickly going beyond himself, and he leans forward to emphasize that "it's also important from the lens of healing, of being able to heal, not just for myself, but for my family, my community, and for generations, seven generations back and seven forward."

With these words of including everyone, Edgar states his conclusion that his healing is not his alone, and he reinforces this as he goes on, talking about long-term healing in the community. Just as the violations are not his alone, like he and his playmates playing in the toxic runoff from the fields, neither is the healing.

"After almost a whole decade of incarceration, it's going to take twenty years to achieve balance, to get to a better place. Again, it's particularly important because the messaging that we receive in there, from being a young adult all the way until those doors open again, is that we do not mean anything. We're called by a number, with the emphasis of devaluing you, not a lifting of who you are, but a stripping you away from that. And you get accustomed to it. I remember being out and looking at my license and then memorizing the whole number. Then I thought, 'Who does that? I don't have to memorize these things any more. I'm not this number.'"

He pauses, then his face lights up in gratitude, again going back to the balance, an infectious attitude and smile going throughout all his words. "But within all that madness, there's always something good that happens, and there are always good people that you meet, and I met some good, good people. I was introduced to the traditional ways, the Indigenous ways of healing, of sitting around the fire, the talking circle, speaking with those feathers.

"I walked into my first talking circle and this Uncle sat there and looked at me, giving me the rundown, looking at me with, 'Hey, are you Indian?' And I looked at him with, "Oh hell no, I'm not. I'm Mexican.' And he looked at me, 'Oh, you're Indian. Go ask your grandma, and go ask your mom where you guys come from.'

"Sure enough, on my phone date, I called my mother and I asked about where we come from, 'Who's your grandma?' Then she asked me, 'Why are you asking all of this?' And I said, 'It's because they told me I was Indian and I said, no.'

"But she said, 'Yeah, mama Chuey was a full-blooded Indian, you know? Then the Indian from my great-grandpa's side, their blood, comes from the Yoeme, the Indigenous, and they spoke our Indian language. I didn't understand them,' she said, and I said, 'So you're telling me we're Indians?' and she said 'Yeah, no shit.'" Edgar laughs, then brings in the larger view that this once hidden knowledge began in him, of his family, of his roots, and of himself. He remembers the stories of his family moving, and talks about the different tribal peoples of his background, then goes into the values.

"That got me going into understanding all the cultural roots of not just Mexico, but of California, and specifically the Miwok tribe up there in Jackson, where we were then. I began taking in the knowledge and stories of their people, of our cousins up here.

"But I was a knucklehead at that time, and I wasn't listening a lot, so I ended up doing about four to five years in solitary confinement. And throughout the whole time my Uncle Dennis, my spiritual Uncle, would come through and talk to me. He'd bring me songs, bring me books, and share stories with me. And that helped me through, the whole time.

"As a kid, going through solitary confinement, it really threw me for a spin. As a young kid going through puberty, going through an overall change, being in twenty-three-hour lockdown, with one hour outside, it really messed me up. If it wasn't for those Indigenous ways, those songs, and the stories, I would have completely lost my mind."

Talking about the Miwok Uncle who came to him brings a deep smile to his face, and when he finishes, he looks out the window into the light silently, then back to us. "I saw other kids lose their minds, you know." There is another long pause, and I can imagine the Uncle moving slowly through the prison, step by step and person by person, holding the long view of his work.

"Finally I was released June of 2017. I made it out." Edgar chronicles parts of his life then, a job loading produce, a broken wrist laying him off, going to school and not asking for help when he needed it, out of pride and independence, being behind in technological skills, then finally landing at another place that would bring him into his community in a good way, an Indigenous way.

"I got involved with the Milpa Collective, and it's pretty crazy because I met Juan, the executive director, and Eli, a program manager, when I was in juvenile hall at thirteen years old. So being there at Milpa was a kind of 'meant to be' type of thing. And now I think about where we are today. I made it, I got accepted to UC Davis this year. I got off parole this year."

He's proud of what he's accomplishing with his life now, and he uses the word "we" to express it, because it's not about him alone but all those people he's talked about and everything around him too. He hasn't accomplished this alone. Then he quickly adds, "And here we all are with everything that's happening, not just the pandemic, but everything that has been taking place prior to that, and now the major fires that are happening. And it's not normal that this is all happening, none of it. All this stuff is man-made."

Edgar includes it all again, past and present, one person and all people, then reflects on a teaching he got from another Uncle. He tells it simply, almost as though we're older children being reminded of what we should already know, not talking down to anyone, just sharing.

"One of our elders told us there's a healthy fire and a toxic fire. Toxic fire is one that without proper care and proper attention, it goes crazy, burning everything uncontrollably. But when you have a fire that's well taken care of, what does it do for us? It cooks food. It helps warm the house. It takes care of stuff that needs to get burnt out. Sometimes things need to be refreshed through a burning process.

"And there's a healthy fire that burns inside each and every one of us. We can throw in a whole bunch of gasoline, and what does it do? It blows up, right? And then it goes down. But if you feed a little bit of wood here and there, seeing that it might need a little bit on this side or that, it'll be warm and perfect, keeping us warm without hurting anybody.

"And the toxic fire is what's happening. As a society we've been moving from one place to the other being very detached. Everything is a resource. Everything is expendable. We're cruel to one another, being able to consider each other 'trash.' When we're able to do that, unconsciously, to each other, then we'll do that to anything we don't even think is alive."

Edgar holds his arms out to circle the Earth, the sky, to include all living things. One man's life is all of life. This is his purpose in this telling.

"I remember growing up, and recall that littering was a normalized thing for us kids, and this mentality stemmed from the idea of 'You only live once,' YOLO, right? You have a whole bunch of kids saying 'YOLO.' And it's like, 'Nah, you only live once,' so you just throw it down. And that's the individualistic way of thinking at its core." He pronounces "individualist" so the word stands out alone. He puts a demand on his communities and all others. "But what about the kids? What about the next seven generations? What do they get? What do they inherit? Do you know? You're sixteen years old, and you have a kid, so what do you want that kid to have to decide? What they grow up with is bad enough. We all have asthma because we're sprayed with pesticides. Our schools are next to fields where they're used all the time, and then we knowingly throw stuff down on the ground?! Once again, it's not valuing who you are and where you are. It's having that fatalistic mind-set of 'Man, I'm going to die. Why do I gotta do all this?'"

He speaks quickly now, weaving from the fatalism and its source into the most important beginning of change for him, "And so it's going to start with uplifting who you are. Adding value to you as a person is the most important thing that you can ever do because if you know your own value, then you know the value that surrounds you. And when we're talking about the question of how we got here today, it's been five hundred years of devaluing people. It's probably even more, since the Inquisition, and we have seen what religions have done throughout Europe, killing folks who practiced traditional ways of life, and we call them pagan religions. They burned and killed women.

They devalued people, and that same mind-set came over here, and when they saw us, they said, 'You're a workforce. Your simplistic way of living does not merit our definition of civilization, so we will use you through enslavement and now today, cheap labor.' Think of cutting people's arms or legs off for taking something to eat! You're really going to maim a person?

"It impacted generations, and it's still impacting us today. This is over five hundred years ago, and it's still going on today. What came with them was murder, rape, and robbery. And self-hate grew out of that rape, for those men that had to witness partners, community members, daughters, children, wives being raped and murdered in front of them. This created a deep sense of helplessness that was passed down."

Edgar sees it around him, looking back and all around, and in his family.

"My grandmother said that one of the biggest things that impacted my grandfather was witnessing his cousin being killed in front of him. He had to hide in the bush and watch his cousin get killed. And he had a sense of shame, regret, of being not worthy, of being helpless, and all that comes with it, and the survivor's guilt. All this interacts, and it's all about devaluing people and yourself.

"This is what brings us to where we are today, to being able to extract resources from the Earth willingly. The silver and gold that went to Europe came from this side of the world. None of it stayed here, and it funded wars in Europe that killed many people. And then there's the slave trade! It all boggles my mind."

Edgar is looking over his shoulder into the past, then turning forward into today, a gathering of meaning he insists over and over that we look at, bringing it all into one moment, a stream of meaning and history that all people should contemplate. It's a long sense of time.

Then he gives an example of the shorter sense, the truncated history that is colonialism. A city manager in a local town questions whether we should have an "Indigenous Day" because Indigenous people "colonized and owned slaves as well."

"That he is even trying to make sense of it bothers me because that town is 30 to 40 percent Indigenous. You have a big community there, and a city official *questions* whether we should be honoring this history and these people?"

It's the questioning that bothers Edgar, so he addresses the anger that rises out of that short view.

"It's in everybody's psyche, 'Why should we care about anything if nobody cares about us?' That's the crazy way of thinking about it, crazy, but it's from being just a survivor, only thinking about your own well-being, your survival by any means necessary. If that means throwing that person under the bus, you're going to do it. If that means you do crazy stuff, you're going to do it. And the results have a direct impact on the world, on everything around us.

"In the Salinas Valley, we have an agricultural business that produces upwards of $5.6 billion a year. How much of that food stays here? Everything is sent out. We have some of the worst water, air quality, and living conditions, some of the worst in the whole nation right here in our own backyard. And we see the impacts of climate change. We don't see the monarch butterflies any more. They used to be all over. I remember watching them as kids, from Pacific Grove all the way to Santa Cruz, but you don't see them anymore. And again, all these fires!

"It doesn't just happen, right? Five hundred years of straight, deep colonization has ultimately impacted how we look at the land and how we look at people only as a means to produce something."

He stops talking, looking directly at us, and the frustration is palpable, the dismay rising between the three of us. There is a long silence, then I reflect with Edgar about the fatalism he describes, how it's handed down from generation to generation "to everybody, along the tracks, along the levee, and everywhere else," I say, "but the Miwok elders put a spark in you that made you sing the old songs today. You didn't lose your mind." He shakes his head and nods at the same time.

I ask him what he thinks about how all that he is describing culminates in a radical impact on Earth and all life, and I ask, "In the face of it all, and putting it all together, I want to ask you, 'What's your Indigenous dream?'" He laughs aloud at my use of the words but not in the least at the question.

"I've thought about this because in every organization you have to have your vision and your mission, all your big thinking. For me, it's always been the reclaiming of our land and of our language. Our language gives us different perspectives on how we see the world, on how we're able to name things. Moving to a language that is more poetic, more fluid, more of just acknowledging things, is a whole different level. My dream is of reclaiming that language for all Indigenous people, reclaiming their own languages, traditions, and cultures.

"Land is also a big component because it allows us to practice our sovereignty. Land allows us to practice our traditional ways of farming without governmental constraints. Harvesting the value of each other also includes having our own educational systems. That's how we build the narrative. That's how we go back to our people. That is how we message to them right now." He describes how "everybody has already established their message for the next twenty, thirty years, forty years. We need to do that for our folks too. We need to do this for all our Indigenous communities.

"I think back to what the American Indian Movement did, and my demands today would be the teaching of our languages within the public education system. And if not there, we'll open up our own educational systems to teach folks all the different languages that are indigenous to this land. Once we can see and speak through our own language, it's a whole different ball game. How you identify will be different. You told me Shannon Rivers said, 'I'm not Native American. I'm not Indian. I'm something way older than those things.' That is *very* fine!

"There are many things, like a reclaiming of our foods, a reclaiming of our own way of life, and our own means to our own health care system. We need to have our own means to produce and distribute. We need our own stores because those are the ways we bring the foods that we cultivate to everybody in a community. But again, it starts with language because once we have a grasp of that language, it will impact everything else.

"For example, in a class I'm taking now, two elders from Mexico came, and we asked them, 'How do you say this?' [They answered,] 'Well, we don't have a word for it, but we can make a word for it right now, if you want.'

Maybe an English or Spanish word for our question doesn't even exist in their language, but if you ask them how to say 'love,' they can tell you how to say love in many different ways, how to say family, how to say brother and sister, in many different ways. And it's not the same thing all the time but something different. They said, 'We don't know how to, there's no word for that, but we can make one up right now!' That's the Indigenous dream, being able to speak to our tree relatives, our plant relatives, the animal relatives, to all these things, all these different relatives that we have, to speak to them in a language that is almost as old as them, almost, because they're older, and they can recognize us, and they can understand us. That puts us back there once again, throws us into the mix, back into the greater circle of things.

"We always say at the Milpa, 'We're trying to break cycles and rebuild our circles.' It's the cycles of incarceration, abuse, and violence that we're trying to break, and we're trying to mend our circles, those circles that were important to our community before and were so powerful. That's my Indigenous dream, and it's very simple.

"And we can get to the younger people and start empowering them to speak. We have a lot of Indigenous people from Mexico coming over here, working in the fields, and they get ridiculed. They get mocked, made fun of. 'You don't talk Spanish, you don't talk English. Then what do you talk? You're not useful.' We need to be able to allow them to feel comfortable and empower them to talk. We had a fourth-grade Indigenous child in our office at Milpa, and he opened and closed the talking circle for us in his language, and it was the most beautiful thing I have ever seen. That's what we need.

"The scientists and theorists have come up with a lot of new things, but the deeper teachings have always been there. So it's putting our ego to the side and listening to others and their solutions and their teachings, regardless. Having or not having a degree is not going to be important, but listening to those elders, to those grandmas, is important because they have everything. They're the wisdom keepers, and listening to them is important."

Edgar stops to laugh at this point. "Sometimes I get yelled at by them, but that's okay. I apologize for laughing, but once I made the horrible mistake of trying to go into the kitchen when they were all in the kitchen. When people are eating, I like to go grab some food, but that was a mistake. I almost got jumped. 'Oh, okay. I'm sorry!' I *ran* out of there!"

We stop to think about all that's been said so far, and we recognize the power of the grandmothers who have been so silenced by the system of thought put upon us all.

"I don't know if I'm answering much for y'all. I'm just witnessing what I'm seeing around us, like the fact that on Saturday there were three shootings in Watsonville. Two young men died. One I knew. Even amongst all this craziness that's happening to the Earth and everything else, three men were shot. I think, 'You're supposed to be home with your family taking care of business, and we're already at risk of dying, and you're out there just making it faster!' It makes me know all the work that needs to be done, all the healing that needs to be done, and I know we have limited time."

Edgar lists the physical signs again, the hot and "dry as toast" October with no rain in sight, the empty reservoirs, and the vanishing wildlife. He points to the coastal migratory pathway of birds that's almost empty. "And you think, 'What the hell is going on?'"

A long silence comes, looking at the inevitable consequences of colonial thinking and behavior, as he describes it, then he says, "We're at a moment in time that allows us to really reflect, and hopefully once this door is open, people will have a different idea of how to navigate, how to appreciate things. You know, people were appreciating toilet paper in March and April, so I hope they really appreciate it when it comes down to seeing how we get to where we should be."

There is another long pause, and I think about the Miwok elder telling Edgar to ask his mother who he is, and the humor Edgar has in telling us about that, and the appreciation of toilet paper, and I think of the ceremony in the prison. I ask him, "Can you talk about the place of ceremony?" Edgar's face lights up, in seriousness.

"Two weeks ago, a young man got out of incarceration and came asking for ceremony. I haven't had ceremony since I've been in quarantine, but I know that we had to connect him to that fire and water, take him back to that lodge where he can talk to those ancestors, those stones, to those elders who talk to that fire and that water.

"When he goes back to the neighborhood, they're going to give him a pistol and some dope. 'Here, protect yourself. And here's some dope, make some money, so you can get on your feet.' I know this because that was me. I was in that same position. They look at it as a way of helping you out, you know? And that's the mind-set that runs rampant in our community. You ask them, 'Man, what are you going to do this weekend?' 'I don't know. I'm just taking it day by day.' I ask people, 'Where do you want to be a year from now?' 'I don't know. I take it day by day.'

"Hearing that over and over again is sad. They don't think they're going to make it. They understand and accept the reality of being dead or incarcerated. And they would rather just live their life to the fullest as much as they can.

"But when I get them on their own and I'm able to talk to them, I always make it a point to make sure to tell them that I love and appreciate them. And they know that I'm no longer out there doing what I was doing when I was younger.

"They know that, and they say, 'Hey, here comes this guy with this Indian stuff.' But I tell them, 'Show me a picture of your grandma. Show me a picture of your grandpa. Are you going to tell me they're not Indigenous?' 'Yeah, bro, but we weren't raised like that.'

"Yeah, we were. We were. They have the reservation. We have the barrios. It's just that we have to go back to how it was. We have to reclaim what it was back then and reclaim who we are as a people. We have to bring that young man back into that fire, talking to that elder. He's at risk of being killed, of falling victim and not having a clear sense of direction.

"So being able to reintroduce our young people back into these ways of life is important. We take them to the Mexica New Year in San Jose, with them asking questions like, 'Man, what are they doing?' And we say, 'Hey, do you want to go ask them yourself?' And then the little kids go ask if they can go inside the circle, if they can dance too. And having that for them is beautiful.

"But it's really hard because the trauma is baggage that our folks carry. They might have their head high and their chest out, but they're dragging, they're sinking. They are sinking in their own sorrow, drowning in it, and in their shame.

"But once you bring them close to that fire and they feel the warmth of that sweat lodge, they see the feathers, they smell that cedar, they smell that copal, and the sage, and the bear root, it awakens something in every single cell of their body that allows it. 'I know this from somewhere. I know this. My grandma used to burn that. You know what? My grandpa talked about something like that. You know what? My grandpa had feathers too. And he had a box like that too. He had a cedar box. We never knew what it was for.' So it's been there. It's just hasn't been nurtured. And as they keep getting closer and closer, these traditional ways are what's going to help us reclaim our neighborhoods."

Dahr has been listening silently, deeply moved, then he says, "It's an honor hearing you, and thank you for being so open." These are words of a man dealing with what he saw in the Iraq war, so he knows the difficulty of speaking. "Can you talk about why your revelations took such extreme measures?"

"It took this long because I was still finding the value of who I am. I was battling with myself. I knew what I was capable of, and I had people that were telling me this, like my mom, but I didn't really listen. But once I saw that I was capable of actually doing good things and great things, that I was actually something, then I found that component of being proud of who I am and where I come from. And this didn't just happen.

"As soon as I went into a sweat lodge, it began a process. It changed my perspective on things, but when it really hit home for me was two years ago after a ceremony, and I was able to really understand why it took me this long.

"It was about generational healing. That medicine that I was receiving in that lodge wasn't just for me. It was for my grandma who never hugged my mother. It was for my grandma's grandma who left when they were kids, and for my grandpa, from when he got killed. It was for when my father was a kid, for my father who was incarcerated and never dealt with it. It was the healing for them and us, but it was also the healing for my sisters and my brother, with my sisters having to watch my mom go through all the shit, from heart problems to diabetes to being in an immigrant detention center, risking getting deported, having to see her go through that, and me missing that.

"That's when I finally realized why it took me this long, understanding that it has never been about me. I'm just a small part of everything that's going on. And the healing that we bring through going into these Indigenous ways, learning even just a little, like just how to say good morning, how to say my name, to introduce myself, speaks volumes because now other people can hear me.

"Even my sisters who make fun of me hear me. They say, 'Man, you're tripping,' and I say, 'No, this is the way.' And when something hard goes down, they ask, 'Hey, can you bring a little bit of cedar? Can you bring some copal? I've been feeling off, so can you fan me off with your fan?'"

Edgar goes back to the enormity of the realization that connected him to everyone around him, and he says. "It became a matter of life or death. And this Indigenous way is the only way that I felt that my prayers were heard. When I spoke with that feather and I smelled that sage, that completed something ancestral deep inside me. The elders reminded me that the eagle feather took my prayers where they were supposed to go. I had not been able to understand why it took that extreme to get me here, because I was just thinking about me, me, me, and once I was able to snap out of that and see the generational connections, I was able to go to someone else.

"Grandmother once told me, 'You always hear people praying for themselves, always praying for themselves. It's always, 'God have pity on me. I'm just a pitiful man.' She said, 'I don't pray for myself. I pray for everyone else, because I know that someone's praying for me already. So I just make sure that I pass it around.' I thought, 'Damn!' and this is one of the grandmas that yelled at me to get out of her kitchen."

We all laugh, then he adds, "There are different layers of these things, and it's important to know that when there's balance in the family, there's balance in the community, and when there's balance in the community, we're balanced with the environment around us."

The interview is complete. We smile and relax, then after a while of talking and laughing, we share our thoughts about the manner of our approach, of letting it roll, letting him say what he needs to say. We ask for his thoughts.

"You know, I really wanted to do this because the opportunity to get to talk to different folks is important. Being able to relate a message is important. Books continue to be the number one way of how folks are getting information. And if we can get this message to folks as soon as we can, it's best. I think they need that hope. Folks need hope."

As for our method, he says emphatically, "It's important to go back to this way. There's a dicho in Spanish that says that by talking you fix things. And this choice goes way back, because back then in the old days you could go to anybody's house, like when I would go to my friend's house and his mom would sit me down to eat some food, and while we're eating food, they would interview me, ask me about my mom, where my mom comes from, where everybody comes from. We're basically breaking it down. And so the way y'all are doing it is extremely important, because I've been in those kinds of interviews where they've asked me the questions and, 'Oh yeah. Thank you.' Boom. Next thing, I look at the interview and ask, 'What happened here?' And I've had to reach back out. It's like I'm operating under their norms, their way of working, but here we're operating under different ways. You've introduced the respect for the story, the respect for the individual, that respect of actually going back to those traditional ways. So for me, it's important and I appreciate this way of doing it. A lot of folks will see it, and I think there's no better way than to do it like this.

"When we do our interviews for a job in our internships at Milpa, people come in and this is what we do: 'Tell us about yourself, everything.' We ask them, 'Who's your mom?'

"'What does my mom have to do with it?' 'No, it's important because your mom tells us a lot of things, you know. Can we meet your mom?' 'Why do you want to meet my mom?' 'Yeah, we want to meet your mom, because my family's here, and your mom's going to give us the scoop. Your mom's going to give us the fullest, so we don't need a resume. Just bring your mom.'

"So what you are doing, by handling this like this, that's basically what you're doing. That's how I look at it. So I appreciate y'all for heading that way.

"And, I can talk," he adds, laughing.

12

Alexii Sigona (*Amah Mutsun*)
Stewardship

COMPOSED BY DAHR JAMAIL

We can't just do a ceremony on a parking lot.

—Alexii Sigona

Currently a candidate for a PhD in environmental science focusing on Indigenous land conservation and food sovereignty at the University of California, Berkeley, Alexii Sigona is working toward having the letters after his name so that he can help his tribal band, the Amah Mutsun, protect and steward their sacred land.

The stated Conservation Mission of the tribal band is:

We are Amah Mutsun, of the lands known to us as Popeloutchom.
Home to our four-legged, winged, finned, and plant kin;
they have provided us with all that we needed for millennia—we will
   care for them.
Resting place of those that came before us and cradle of those yet to come,
they are sacred—we will protect them.

The heart of their ancestral lands is a place they call Juristac (pronounced Huris-tak), in the southern foothills of the Santa Cruz Mountains in California. Mutsun ancestors lived on and conducted their sacred ceremonies there for thousands of years, but now an investor group out of San Diego owns the land and aims to develop a 320-acre open pit sand and gravel mine there, which the Amah Mutsun Tribal Band vehemently opposes.

Alexii explained to us that Juristac has long been known to be a place of power, a place where spirits lived. He told of how their ancestral lands house what they call "the spring of eternal life," and of how it contains volcanic tar pits which are one of the central components of Juristac. He spoke to the "silliness" of the idea of drawing a boundary around Juristac, and wonders how one could place any kind of constraint around the Amah Mutsun's powerful spiritual being known as Kuksui, let alone their cultural heritage.

Juristac translates to "place of the Big Head," and is where Big Head dances connected with Kuksui and other healing ceremonies that took place for centuries, if not millennia. The area is full of critically important cultural sites and Earth features that are of significant spiritual importance for the tribal band.

"There are stories about the spirits becoming Big Heads because of the power of the place," Alexii explained. "These were related to earthquakes, as the Big Head would appear because of the rumble. So in that sense, you need that site to do those ceremonies. You can't just move a Big Head ceremony somewhere else, because that's not where the Kuksui spirits are. The Kuksui spirits live in those hills, live on the coast, live in the oak trees and fly down."

Kuksui is a spirit that, according to Alexii, has a lot of power, and is also a trickster. It is a spirit not often discussed, because people are wary of getting on its bad side. Because of this, there is not a lot of recorded history of people talking about spirits, or these ceremonies, because of their power.

Alexii also pointed out how the collection of medicinal plants, along with things like mugwort and tobacco, have the power they have precisely *because* of the place they are gathered.

"Although you could gather mugwort somewhere else in Mutsun territory, you'd want to gather it there [Juristac] because it has such important power with those things imbued in it. The spring of eternal life is right there. Those volcanic tar pits are right there. The medicine men are living in the hills and the spirits are there too. So all these things make up this really important cultural landscape that makes things like medicine have power, and just being on a place of power is good for you."

But the tribe does not own any land within what was their traditional territory. They are not "federally recognized" despite a long and well-documented history, so like other such peoples, they have no official land base. Alexii sees what they are fighting against today as little more than a continuation of the colonial violence his tribal band has endured since first contact.

Despite Juristac being "the place of my grandparents, they were kicked off during the lead-up to the attempt to develop the land" during the colonial era, he explained. One of his family members continued to return to the site to gather medicine until the 1920s, but since then they haven't been allowed to do so by the "owners" of the land.

Nevertheless, the tribal band continues its work to protect the land.

"We have to say it's our most sacred site," Alexii said with urgency in his voice. "We have to try to explain what it is to our people and how it qualifies us for the National Registry of Historic Places."

What troubles him is that in order to protect Juristac, the only way to access it and be stewards, is, as he told us, "We're going to have to own it."

Many of the Juristac campaign partners working with the Amah Mutsun to save the land are wealthy Bay Area environmentalists whose money comes from Silicon Valley. Given the amount of land destroyed and contaminated to acquire rare metals used in the production of technological devices, Alexii is faced with a dilemma he openly confronts.

"It's like we're using the profits of someone else's sacred land being mined to help protect our sacred land. Like being intertwined in the Bay Area politics that we have to play, which requires this sort of activism, this is difficult to think about. The outcome, regardless of what happens, is going to leave me with some elements that are a bit unsettling."

Regardless, Alexii wishes Juristac could "just survive on its own," and laments the fact that the sacred place does not have the right to thrive, and referred to these vexing problems and dilemmas as causing "a split in the relationship with our traditional territory."

"This is a really clear case where we have these beautiful relationships with our traditional territory, and there were forms of governance there, forms of ceremonies that were taking place for hundreds and thousands of years."

By way of example, Alexii discussed how, within tribes, people followed rules around how to relate with, engage, and steward the lands where they lived.

"Native plants that were important for us were important because they had an essence that was important, but also because we managed it and were the stewards of it. So when I see a really important cultural plant growing somewhere that I've never been, I don't necessarily think it's the same type of relationship or importance as in the place that I've been from, because I've not stewarded that plant in that particular location."

This is a difference that Alexii believes needs to be addressed, in order for people to care for and steward land properly, yet also behave with equal respect to places they are not stewards of.

Alexii discussed differences between the Spanish and Mexican periods of occupation of the region, and pointed out the important element of the American period being the introduction of fences and a broadening and deepening of the concept of private property.

"This is not yours," Alexii said of the core message of having a fence around land. The privatization of land enabled the government to kick Native people off lands where they lived, and gave people "the right to exclude." This became part of the colonialist culture. Alexii sees this as the worst thing that could have happened to Juristac because his people could no longer live there, and once the tribe was prevented from accessing Juristac, they could no longer conduct their ceremonies.

Hearing this made me wonder about the now ubiquitous lack of connection to land around the world, as the dominant culture has been writ large with the concept of private property. This prevents the connection to the land physically, psychologically, and spiritually. It seems to me this is the fundamental cause of the climate crisis, and actually made it inevitable once privatization became nearly ubiquitous.

I asked Alexii to share his thoughts on this.

Using Juristac as example, he talked about how his ancestors created things from the land like flutes, whereas when ranchers privatized the land they put oil wells on it in order to make money.

"Before the land was privatized, the mind-set wasn't skewed toward production of materials and monetary gain. You wouldn't have everyone being so gung ho about producing oil. Instead, you would be thinking about what is best for everyone."

Underlying that thinking, Alexii went on to talk about relationship.

"One component that I touched on earlier was about relationships with our plants. The relationship itself, making those plants important, making the relationship itself significant, generates a kind of positive feedback loop."

He wonders about how, without that active relationship, anyone could truly be a steward of the land where they live.

"If you're tending a certain plant population for the past four decades, you may notice a change corresponding to climate change in the ecosystem. And if you notice that, then maybe your management systems would change slightly. You'd adapt, and that relationship would be able to let you carry on stewarding it in a good way."

Instead, he sees the insidiousness of the thinking that comes with private property seeping its way into too many areas. Even in public land systems and national parks, exclusion and disconnection from lands continues.

"There's a way of having a hierarchy and asking permission which also inhibits our way of kindling that relationship. Even by going out for scientific purposes to gather data, or make measurements, you are just doing that and then walking away. So those systems in place are not conducive to us having good relationships with these lands. And what I'm worried about is that I think our society just needs to have better relationships with native plants, or even just with our natural ecosystems."

Alexii thinks that even if, for example, all of the undeveloped coastline in the northern area of Santa Cruz County is restored to native grasslands and protected, there aren't many people that have relationships to that space who know about it and who can understand when it changes.

"That's not going to help us move forward. It's still going to be something in the background that someone takes pictures of for their graduation, or something like that. They're not going to be able to visit it. They're going to have to stay on the trail. It's going to be cordoned off to most people."

Another obstructing factor is accessibility, or lack thereof, due to having signage and information only in English. He sees this as exclusionary of people who live in the area who don't speak the language of the dominant culture.

Alexii sees everything we are dealing with now on these issues as stemming from the philosophy of assumption of ownership that began only 150 years ago in his region.

This, then, as he sees it, has expanded out into causing humans' disconnection from the Earth all around the planet ever since. And if we do not feel a deep connection to the place where we live, that which sustains our very lives, we will not take care of it.

Stan brought up how the rule of law is assumed to be based on moral principle. He asked Alexii what the moral principle was underlying the assumption of ownership in the 1850s, and, "How does that apply to what the Amah Mutsun are confronting today, given that we are all having to adapt to a system that has a very questionable moral underpinning?"

Alexii pointed out how, during that time, treaties with the Indigenous in his region were being signed, and his people were supposed to be given land in California's Central Valley as part of a 7.5-million-acre reservation.

He went on to ask if it is a strategy "to obtain justice by pointing out past injustices, then asking for reparations. Is that what you're supposed to do for justice? Show the United States how they err in their own legal system, and that they are wrong and contradicted themselves, in order to get justice?"

Beginning in 1968, people were compensated for not getting the land they were promised in the Central Valley for the reservation, but were paid only a minor amount per acre for it. "This happened," Alexii added, "regardless of the United States kind of saying, 'This is not yours forever' and dispossessing people from their land."

Despite their dispossession from the land and lack of federal recognition, members of the tribal band continued to do what they could to maintain their connection to the land, and pass down the stories of having done so. Alexii believes it is important that these family stories continue to be shared, as it is also a form of resistance to what has been done to them.

"Although the U.S. government is trying to impose these private property norms on it, I go to Juristac regularly and trespass. I gather medicine. I've been able to have a good relationship with that land."

Alexii regularly prays and offers tobacco for the land when called to do so. "It's a really powerful place," Alexii said of Juristac.

In addition to the aforementioned, another thing he does that he sees as a form of resistance is flying a drone over the area to access it, and using it to search for more of the culturally important sites.

"We don't know where the spring of eternal life is that is in our records. We don't know where all the volcanic tar pits are."

He is grateful Juristac hasn't been developed like the rest of the Bay Area, which would have utterly destroyed the area.

While a key portion of Juristac was recently sold off to a development company, the tribal band doesn't have all the details on the transaction, in what he described as a "sketchy" transaction where the owners are "just trying to make more money off of it."

So time, as ever, is against their efforts to protect what remains of the area.

The implications of private property and the inherent message it carries of "This is not yours" are deep. I asked Alexii if he would talk more about the philosophy, and how it impedes people's willingness to take responsibility for land in order to care for it and be in right relationship with it.

Alexii sees the idea of private property akin to having "the right to exclude." He acknowledges the bind the Amah Mutsun face. "The closest things we have to property rights are rights to exclude through memoranda of understanding, conservation easements, and other agreements with public park or state agencies. We do have property rights in that sense, and with those, the right to exclude other people and to have these practices and ceremonies on top of the hill where our creation story took place, without other people around."

Yet, they still have to ask permission from an agency, like California State Parks, for example, to exercise those rights in order to hold a ceremony within Juristac without other people around.

"But is that enough for us? Is that what we need? Or do we need to own it? Do we want to be able to exclude people when we want, or be the ones in charge of directing the stewardship and everyone having to come to us if they want to steward or access it or something like that?"

Alexii shared how Amah Mutsun tribal chairman Val Lopez consistently reminds them they have access to 140,000 acres of their traditional territory, but pointed out that this number comes from adding up all of the memorandums of understandings (MOUs) the tribe has with county, state, and national parks.

"But no way in hell there's people going out to all of those places," Alexii said.

His point was that he does not feel like his people benefit from the MOUs, and he sees access rights as "the ability to benefit from things. When you think about access, it's not just about private property. You cannot own land and benefit from it. You can have access rights like we do, where we're able to gather, where we're able to hold ceremony, we're able to exclude other people. Those are all access rights. Those are not about responsibility."

Alexii believes private property wreaks havoc on all of this, and cites the concept of national parks as example. "If we have a sacred site and it's owned by the public, we could probably get out there and access it and have a certain amount of access rights to it. But if it's private property and it's a sacred spot, then we don't have access rights to it. We're not able to benefit from it. In addition, because of how easy it is to do a sand and gravel mine, for example, on private property, we lose our ability to benefit from it for future generations, in addition to the benefit of just knowing that we have a sacred place, even if we can't occupy that space, but can still access it."

Thus, with the ability to own comes the ability to deny the benefits, exploitation included, that come from the land to other people.

"We have loads of culturally significant places in Amah Mutsun territory, and some of them are owned by private property folks. And we have to mount all these campaigns just to keep sites from being destroyed. And that is a form of stewardship. And then on the other side is when there is an important site in state parks. And so what we have to do is we advocate for us stewarding it, and we steward it and we protect it and we actually do conservation work on it."

So modern stewardship for Alexii and the Amah Mutsun means stewarding land one way if it is privately held, while if it is held by the federal government via scientific permits, then they must steward it another way, whereas with state parks they have more rights and leeway.

Out of obvious necessity, these are fragmented and disjointed approaches. The tribe nevertheless is doing everything it can to safeguard and live in relationship with the land of its origins.

Nevertheless, again, they are not allowed to live on the land, which has been their place on Earth for millennia.

Alexii describes his PhD work in environmental and social science more like political ecology. He does so because he's looking more at access and collaborative conservation, along with food sovereignty issues.

From that frame, Stan asked if Alexii would discuss whether he sees other Native people in school doing work aimed at addressing the myriad issues confronting tribes and access/ownership/private property issues within their homelands, all of which is directly tied into the climate crisis, and if he found hope from that.

Alexii is grateful for being in academia, and having the ability to pay the bills while doing work that will ultimately serve his tribe. His goal after he completes his PhD is to use his doctorate to work toward helping his people get more rights, and amplify the marginalized voices in his community that deserve to be heard and listened to.

"So that's my role. I think it's difficult being a student during this time because a PhD is the long game. It feels like you're in here for a while and you just have to learn, and you're still learning. And I feel like I don't know a lot. I really don't think I have answers right now. And I don't think I'm going to have answers for a long time."

This was a challenge for Alexii, as he watched the catastrophic wildfires across the West while he was in the midst of his learning and preparing himself to be of service after he leaves academia. Like so many college students during this time, Alexii is also worried about whether his degree would even be useful, "because there're so many things that have happened in the world that make it not important."

He also pointed out other challenges, like the rift in the tribal communities around federal recognition, with some seeing that as the way forward, while others do not.

Due to all of this, he does take heart from knowing of the work of other Native academics, even though there aren't that many. It's not an easy position to think about, with cultural implications.

Alexii is unsettled by how his voice might gain more attention in the tribe simply because he will have a PhD. "Because I'm in an academic program, somehow my voice is recognized as having more validity over other tribal members. And I am getting this degree and these letters behind my name to get my tribe more rights, but it is unsettling sometimes that I might be more listened to because of these letters."

It struck me how his concerns and inner struggles are likely reflective of the complexities and conflicts younger Indigenous people experience in the United States, and likely around the world. Having to operate within the White world, and obtain requisite tools to work within it to try to help their people, was clearly a serious challenge to Alexii.

"Taking care of Mother Earth is a lot of responsibility. It takes a lot of work. But if you're talking to people about your tribe, stewardship, and researching, you're not necessarily the one doing that work. You're working with other people who are doing that work."

Because of that, Alexii wanted to meet with people doing the groundwork, and spend less of his time in public speaking engagements and attending conferences.

"The people who have the sustained relationships with an environment, who do a project longer than the duration of a PhD, those are the people that I'm interested in talking to."

He pointed out the Klamath in Oregon who were doing active fire management on their lands as the kind of folks he wanted to learn from, as his tribe already was. He also singled out Winona LaDuke and her work around the rights of Mother Earth as someone to emulate and learn from.

Meanwhile, Alexii sees the work the Amah Mutsun are doing near Juristac as much a revitalization of the area as it is restoration work.

"The coasts are eroding archeological and cultural sites into the ocean. The fires are way too hot and not indigenous because they're too hot and burning everything up, while development is causing us to lose the ability to steward our native plants in that development site."

Alexii used a social science term, "strategic essentialism," that is now upon us, given the converging and accelerating crises. "Stewarding our plants and talking about protecting our lands as strategic essentialism is important, and it is also a way of talking about ourselves and our community in that manner, by talking about ourselves in the manner of us being the only six hundred people left that know how to take care of our territory."

Alexii pointed this out due to the fact that there are only a few Amah Mutsun left that have the knowledge set needed to properly steward their specific territory.

Alexii is also asking macroperspective questions around what is necessary at this moment in history.

He acknowledges the importance of activism around protecting the shell mounds in Berkeley, and progress made on developers having to gain consent from Indigenous people before making moves on projects. Alexii sees this as an improvement over the mere "consultation" required in previous legislation, which only provides a thirty- to sixty-day window for comments.

But he was already wary of "playing politics" within the state system, having to deal with land developers, and is "still trying to figure out" what it means to be in California's Bay Area and working in terms of conservation and collaborative stewardship.

Alexii does not have an answer for the moral dilemma of wealthy Bay Area nonprofit organizations funded primarily by tech money, which comes at great environmental cost. Nor does he have an answer for the fact that there remains a significant amount of land yet to be developed that could be protected, but how? He believes the Amah Mutsun certainly would be able to revitalize land in his region that needs it, but how does one fit all these pieces together? With this, on top of relatively undeveloped land, or land being challenged by development, there is the question of already damaged lands, and how to respond to them.

"The ability for us to restore our relationship to our environment is predicated upon the availability of native rare ecosystems. We can't just go out to a certain landscape that's been degraded, that's been developed, that's been exploited, and be able to restore these plants, this ecosystem, this land, and ourselves."

"So our cultural survival is dependent on these rare few remaining regions or preserves in our traditional territory that have the ability for us to form these socionatural relationships."

He paused for quite some time, then continued.

"What I worry about is the future generations. If we don't work hard on keeping these culturally significant landscapes tended and okay, open and undeveloped, then what will we have? Maintaining our culture, restoring our ways, having the traditional practices would just be impossible. Then, with climate disruption and sea level rise and the catastrophic wildfires, it puts that even more into question. That is unsettling, and maybe a reason why we need to have more stewardship right now on the ground than ever before in order to get more people to feel that way."

He again paused, this time for longer than before. Alexii looked off into the distance, then back at us. When he spoke, his voice was lower, with determination and resolve.

"That's why places like Juristac are so important. We can't just do a ceremony on a parking lot."

13

Tahnee Henningsen (*Konkow Maidu*)
The Courage to Remember

COMPOSED BY STAN RUSHWORTH

> In order to really heal from something, it has to be acknowledged.
> —Tahnee Henningsen

"People have forgotten that they're family."

These are the first words of Tahnee Henningsen as we begin to talk about climate disruption, the pandemic, nonhuman extinctions, and both environmental and social injustice. Tahnee is a young Konkow Maidu student at the University of California, Santa Cruz, studying psychology with a clear sense of purpose that comes from her life experience. Her aim in joining this project is to share that experience, out of a deep sense of connection and compassion. I had approached Tahnee about participating in this project after witnessing her quiet and deep influence on others in a classroom. She helped a diverse group of young people deal with very difficult issues, through her steady listening and courage.

Tahnee was glad to join, and she continues her opening thought by focusing on her own community. "My reservation is a really small one, not like the bigger ones, and on my reservation, we are quite literally all family. And even there, we have kind of lost that kinship in the community that we once had. We pass each other, even though we have the same blood in our veins, and treat each other like we're strangers." With little hesitation, she speaks firmly of her goals in meeting this problem. "That's something I've been trying to change. I want to rebuild the relationships on my reservation." She has started by encouraging the parents in her immediate and extended family to get children to play together regardless of how the adults might see each other, aiming toward a new generation of family.

Tahnee has a kind demeanor, both soft and strong, and she quickly expands her thoughts on relations, reflecting on the effects of the world surrounding her small community. She is not condemning, but simply observing. "And yes, this is a good reflection of what's going on in society on a larger scale. So I want to rebuild connections with people not only on my reservation, but in general.

"When I walk around in downtown Santa Cruz, or in San Francisco, there's homelessness everywhere, and I have to walk past it and put my head up and ignore it." She leans forward to emphasize her next words. "But it's hard to go past and try to ignore the pain." She goes further, adding that when we do this, "We ignore the suffering of another human being, which could very well be our brother or sister, or it could be us at one point, but we have to stop caring." She shakes her head slightly, then reflects, "I don't know at what point we became desensitized to the suffering of other human beings."

But there is another side to the avoidance, a symptom of a deeper, more pervasive struggle. "At the same time, it's hard for me as a woman to show compassion because I'm also afraid. I have to keep my head down to avoid catcalling and stalking and harassment. I have to present a mean face to the world to protect myself. So it's hard to find the balance between wanting to have real connections with everyone, and also protecting myself and my heart."

Tahnee continues in a measured and calm way, unhurried, and her words are very clearly the result of years of careful thought and observation, as she explains her views of a general disconnect. "It has a lot to do with the way that American society is set up. The American way is very individualistic. It's looking out for yourself." She focuses on how this has affected her people historically, pointing out that "we lost that community idea and lifestyle when we lost our culture, when we were pushed onto reservations. And the land allotment policies also broke up families that were once united. And now, I don't know." She thinks a moment, then explains a crucial fact that remains in her community despite all the attempts to break them apart. "We see ourselves as Konkow, so we do still see ourselves as one," yet there is nothing easy in her observation, and she reemphasizes that her community is not separate from the surrounding world, but is highly influenced by it. "In America, we don't really feel that way. We don't have that connection to each other. Even though we all live under the same country and we are all family in that way, we just don't see each other as that."

Several days before recording this interview, Tahnee spoke about what she wanted to include in how we think about climate change, and that the *emotional climate* is crucial to talk about. She said that the Missing and Murdered Indigenous Women (MMIW) movement is central for her, both the horrific little-known facts underlying it and the beneficial effects of the movement itself. In that discussion, we recalled that when elders are asked how to change all the forms of destruction we're facing, a great many begin with two words: "Respect women."

She returns to this theme now. "When elders say that respecting women is the key to solving the problems, I think of what Darryl Wilson [the northern California Iss/Aw'te writer and teacher] said in *The Morning the Sun Went Down*. When his father took him hunting, he talked about how if you kill one of the younger deer in the front of the line of deer, not only are you killing that deer, but also all the children that it would have had and all of those children's children. As a result, you're killing generations of deer.

"Women are the creators of life, and when the life of a woman is taken, not only are you taking that woman's life, but also all the life that she would have created, her children and their children, and so on, an entire bloodline of people. It becomes thousands of people. It's not one, so crimes against women are truly crimes against all of humanity."

Tahnee gathers this broad perspective with outstretched hands, then brings them together, bringing the feeling back from a broad view into each individual person. "At the same time, the women being hurt there are mothers," she says strongly, waits a moment, then adds, "the mothers of the planet. And this all deepens the cycles of hurt."

She lets that thought sink in, then asks, "Have you ever heard the song 'Daughters'? It says, 'Fathers, be good to your daughters. / Daughters will love like you do. / Girls become lovers who turn into mothers. / So mothers, be good to your daughters too.' The cycle of pain begins with the mistreatment of women. When people hurt women, it affects more than that one woman."

We then discuss the number of missing and murdered women, into the many thousands by now, and the huge number of cases that remain uninvestigated. Seen altogether, Tahnee comments on the effects of it, everywhere and on her personally. "This feeds the fear!"

For Tahnee, talking about the fear of violence toward women is not a matter of being run by the fear. It's a way of seeing all there is to see and to give words to it. In this light, she wants to tell her personal story, including the reasons for telling it.

"What first motivated me to share my own story is realizing that early on I was only able to think of this as something that just happened to me, and I felt pretty lonely." She makes a gesture with her hands, showing a disconnection, then continues, "I didn't relate it to the rest of the world, but now that I'm older, I'm starting to realize that this isn't random. It's a pattern, and it's the reality for Indigenous women. The statistics are just numbers until you become a statistic, and the numbers become somebody that is your loved one. And that's what happened to me."

The words roll from her passionately, and although the sound of her voice comes out easily, the depth of feeling that builds is difficult for her, so she speaks steadily and carefully, following what she needs to say, not only for herself but for all the people she feels might gain from listening. Her motives reach out in the sound of her steady voice.

"In 2008, my mother was in an abusive relationship with my father and she tried to end it. She was able to get out for a little while, probably only a matter of a few weeks, while he was living separately in Sacramento. She, my brother, and I were living in Chico, and he was upset about the whole situation, and was under the influence of drugs when he came in, and he killed her while we were in the house, my brother and I, and later on when the police were chasing him, he ended up taking his own life as well."

She pauses briefly, then continues. "So we lost both parents that night!

"And yes, that always felt like a very personal situation until I started to learn about how often it happens to Indigenous women. And you wouldn't think so, you wouldn't think that until it happens to you. And it isn't random either. It's not like the one-in-eleven-million chance of dying in a plane crash because it's very preventable, and at this point, it's becoming a predicted thing. It wasn't unexpected, is what I'm trying to say.

"My mother was so afraid that she was getting her affairs in order. She was asking people, 'If anything ever happened to me, can you take care of my children?'

"She was trying to convince me and my brother to move to Texas with her, to escape and get away. I think she really knew; she could sense that it was coming. And by asking for somebody to be our godparents, it was screaming for help, telling them that she was afraid, but nobody reached out to help.

"It was almost like she was preparing for the reality. She was making sure that my brother and I would be okay after she was gone, and she was bracing for the impact. It was unexpected for everybody else, but for her, I think part of her always knew that would be her fate, and it was like she just accepted it in the end, and gave up running from it."

Through all these words, Tahnee is fully engaged and connecting with her eyes, not looking away, but now she looks down at her folded hands, then back up in reflection.

"Before she died, I was reading *A Series of Unfortunate Events*. It's a children's book, and in the book, the kids are orphans. Their parents died in a fire, and they're passed around to different relatives. The whole series is about that, and that was the book I was reading that night. The first thing I said when my brother told me my mom had died, was the one thing that an eight-year-old would say when it happens to her. I said, 'I guess the series of unfortunate events can come true.' It was supposed to be a children's book. It was supposed to be a fantasy story, but it was our lives."

Tahnee tries to hold back tears, but she can't, so she allows them, waits, then brings her goal in telling this story to the forefront, gathering herself into her purpose. "What I'm trying to say is that this is a worst-nightmare case scenario. And it's supposed to just be a thing of books, but it isn't, and it's becoming women's realities."

The numbers of unsolved and uninvestigated cases over the years are staggering and sobering, into the many thousands, and the lack of addressing this problem, as well as its roots, is at the heart of why Tahnee wants to include it today. In exact opposition to all other ethnicities' statistics, the vast majority of Indigenous women are assaulted by men outside their own community and ethnicity. There is a residual attitude toward Indigenous peoples that prevails, and Tahnee states it plainly. Her father was not a Native man.

"All around the reservations, there's a lot of animosity toward the Indigenous women."

The roots and effects of the animosity are the larger reality she wants to address, so she begins talking again about how she has learned to put her personal picture into a larger story, repeating that this is for the benefit of those who may be unable to deal with what has happened to them, or what may be continuing in their lives. She connects personal trauma to historical trauma, and says how she feels this should be dealt with.

"I was lucky enough to have plenty of therapy from that point in my life. And for the last decade, it's helped with my personal trauma tremendously. But now that I'm older, I'm starting to realize that I inherited a lot of traits from my mom. I've also been dealing with a serious anxiety disorder. I'll wake up out of my sleep thinking I'm having a heart attack, and debate calling an ambulance, but I have to remind myself that it happens all the time. Even in class when we were watching a film, I started sweating, and my heart was beating so hard I had to excuse myself several times.

"This is hard to understand because at this point in my life, I'm the happiest I've ever been. Mentally, I'm in the best place I've been in my life. I'm really content, despite the fact that I've really lived a stressful life. We all get stressed about work and school and things like that, and me no more than the average person about those things, but when I went to therapy for the deeper anxiety, they just told me, 'Lower your stress.'"

Tahnee chuckles at this advice, laughing at the enormity of that simply stated task, but she took the advice nonetheless, and it led her to making the larger connections that now begin to serve her. Her laughter vanishes as she continues. "I had to do some deep reflection, and I realized that it's coming from a lot deeper in me. It's the effect of what I went through, what my mother went through, and what my people went through. And I'm feeling the echoes of that trauma still."

She expands on these feelings, connecting her own health to that of many others, addressing what she is learning and what it means to her personally.

"I'm a psychology major, and I've been choosing to do a lot of my research on historical trauma. I've been looking at one study of Holocaust survivors as well as survivors of the Tutsi genocide, where they found that the children of the survivors were more likely to develop PTSD, even though the children had never experienced that trauma. They were more susceptible to post-traumatic stress disorder, even without having lived in the time of the Holocaust or the genocide.

"We know that women's health, diet, and everything they go through will affect the outcome of the baby's life. It will affect their height, their weight, their mental health, and how susceptible they are to different diseases. So it makes sense that a mother's mental health will affect the child's mental health.

"That's giving a name, a kind of 'evidence,' for the historical trauma that we already knew was there. We knew we felt it, but now we're starting to find studies and science to back it up, that the trauma is passed down. This is important to recognize because as Native people we have hundreds of years of trauma that still hasn't been addressed. There's a reason why Native people struggle with the highest rates of suicide and anxiety and depression, and so many other health problems. If we recognize early on where it's coming from, and that it's bigger than us, that it goes back further than us as individuals, we can try to understand it."

Tahnee then takes it out into what she maintains is the next step, a crucial step within the larger historical context and surrounding world, a part of the problem that remains to be dealt with throughout the American society she is an integral part of.

"In order to really heal from something, it has to be acknowledged. You can't heal when everybody else is telling you that it doesn't exist. You need the other people to acknowledge what is happening. For the Missing and Murdered Indigenous Women's movement, we need that acknowledgment from the men both within and outside our communities, and also from the police forces that are responsible for looking after these communities. We need that acknowledgment in order to heal." She emphasizes that people must know the facts and the numbers to know the impact on Indigenous women's lives, not with the MMIW movement as an isolated thing, but as part of a larger, much-avoided context.

"It's the same thing with the numbers of the genocide. It's not something that's talked about, especially in California history. In fact, I'm learning things now about my own tribe that I've never known before. The Konkow people had their own Trail of Tears, where we were relocated to a reservation, and a lot of lives were lost on that walk."

In 1863, 461 Native people were force-marched out of their homeland to a distant reservation, and in a matter of only three weeks, only 277 survived. All accounts of the march paint a brutal and outrageous picture rarely taught in California schools, if at all. Tahnee herself is only now learning this, and she's a direct descendant of those very people. "This is just not something that you learn about."

She points out one of the many problems with this picture. "It's kind of like when people lose a family member, or somebody goes missing, and they have accepted that the person has gone. But in order to really get that closure, they need to know how it happened. They need to know what caused it, and they want all the details of it so they can finally close that chapter of that book.

"That's what we're needing. For healing, we need to take back that knowledge, take back that history. And, we need acknowledgment from the government and the people of this country of what happened to our relatives, in order to start that healing process and go into that grief, because that's what it is, grief."

She focuses on the word "grief" strongly, shakes her head gently but firmly, and brings the feeling of it out to a broad understanding of how the emotion works.

"We get it mixed up, thinking that we only really grieve for people who are close to us, but that's not true. Our bodies weren't designed to only grieve for those close to us. It doesn't discriminate what we're grieving for, so that kind of loss is like my own personal kind of loss. It's something we have to process and go through, and we need that information, those numbers, to get that grieving process started."

We talk further about the need for the whole of society to understand its history, how Native American history is American history, not something separate from it. The grief lies in the land, and it permeates all the descendants in one way or another. Some face it, while others turn away from the facts, an unhealthy trait. Tahnee expands on this idea, bringing up the pervasive quality of historical myths and how they are perpetuated in institutions, places where people might not realize the extent of damage they cause over a person's lifetime, beginning with childhood.

"I just did a paper on the Santa Cruz mission, where I was required to compare information from two different websites. The first was the Santa Cruz Holy Cross Church and Mission, and the other was the California State Parks website for the mission. They are so vastly different that it's almost incredible. The Holy Cross website uses language geared toward children, and I think it may use that as an excuse to not address difficult things. Although Native American people are mentioned a handful of times, it mainly focuses on the building and construction of the mission. It's insane. They don't talk about the purpose of the missions. They just talk about how beautiful it must have been, how great it must have been.

They've got nothing of Native American history, and the only thing hanging up on the wall is a silk robe from China.

"It's crazy that it's still ongoing, and we don't realize that this oppression of our culture is still going on today. Holy Cross is trying to ignore the truth. They're not brave enough to admit their wrongdoings, even when right next door, the California State Parks mission is telling the truth and trying to reflect what daily life for the Ohlone people would have been. In this area, there's a lot of work that needs to be done."

Tahnee goes further into the issue of psychological denial. While completely respectful of people's choice of faith, not challenging individual beliefs in the least, Tahnee speaks in no uncertain terms about the omission of honest portrayals to children by institutions. "They're kind of cowards for that denial."

She offers her thoughts on what is possible as an alternative, a continually returning theme for her. "If they acknowledge that at one point they used greed, the desire for land, and slavery, and that they used the name of God to get that, this opens up the question of what else the church could have done. They are trying to keep up the image that all they do is holy, and right. But if they were to say that 'missionaries weren't just trying to convert people to Christianity,' that they were using that as a front for greed, well, things might change."

Change is what she is asking for, through presenting people today with a larger, more honest and analytical picture, again, as she has had to do herself. Without a more realistic picture of real human behavior in a colonial situation, dangerous myths are perpetuated. She adds that "their inability to admit to the truth is showing how it continues today."

In light of the need for truth, Tahnee returns to the atmosphere underlying the MMIW movement, to the fear the crimes perpetuate and how the movement affects her and those young women around her. The omission of Ohlone people's experience on the church website, alongside the silence around her people's Trail of Tears, the simple disappearance of peoples from a key national story, runs parallel to what she says now.

"As women, it's instinctual for us to have that loving caretaking trait about us, but more and more I'm starting to see us retract from our communities to protect ourselves. I see it in myself, and I speak about it with my Indigenous sisters and friends. It's a different world now.

"My partner will go out to the grocery store at night, without a care in the world going to get groceries. That's just how it is for him, but if I'm going somewhere, if I go at night at all, I have pepper spray in my pocket and my keys between my fingers. I'm constantly looking left and right, making sure that there's nobody following me.

"This is preventing me from wanting to move to a bigger city. I want to, but I'm afraid of the crime, and I'm afraid to do things by myself."

Tahnee does not live in the big city now, but in a relatively small university town, yet she echoes what far too many women feel. "It's not fair that I have to have a chaperone every time I go somewhere for me to feel comfortable in my own community. And, I'm starting to see that happen more and more with a lot of us, pulling back and retracting."

In the face of the fear, the MMIW movement has a positive rippling effect for her and her friends. "The Missing and Murdered Indigenous Women's movement gives me a lot of help. For example, the attention that it's brought helped me tell my story for a powwow recently. We were showing a video for the movement, and the organizers asked me to do a segment at the end as a call to action. They didn't know my story, even though I've known these people since I was a kid. And with this encouragement, I finally thought it was a good time to really open up and share. For the first time, I talked about it on video and shared it with people."

She smiles at the memory, not only at the moment of breakthrough for her, but for the sense of community it immediately brought her. As the story emerged, so did the people's feelings. "I had all kinds of people coming up to me and thanking me for sharing my story, and telling me that it really resonated with them.

"One thing that stood out, that I really appreciated, is the men who came up to me and told me that my story touched them, telling me their stories and saying that mine really opened their eyes to what's going on in the community."

She emphasizes the sense of solidarity that began for her with that telling, then returns again to the atmosphere of fear she feels daily. "That's what it is for women in the world right now, even in this little Santa Cruz community where it's supposed to be safe. My partner works downtown and hears stories all the time of women being attacked by random people on the street. It's everywhere."

After this sobering reminder, she continues to explain the importance of MMIW. "The movement is asking for justice and recognition, justice for the lives already lost." She raises her voice and opens her hands, as though asking for and demanding something that should simply be reasonable. "And it's asking for preventative measures to protect women from it happening again."

She speaks of the need for disclosure in the process of personal and social change. "The problem is that you can't really face a trauma that's still ongoing, that's continuing on, so we need more. It's like apologizing for something and just doing it again right afterward. You can't really forgive somebody when there's not that promise that it won't happen again. And so we're really seeking that promise that it's not going to happen again, and asking everyone to face that trauma."

She has no real answers, but speaks again to the need for resolve and confrontation, adding key components for her. "As for facing trauma, I should be an expert, but I'm really not. You face it because you don't really have another choice. It's the only way to heal. You can either sit in your pain or you can face the trauma. But we just need that acknowledgment. We can't be ignored, and with the movement, I don't think we will be ignored for much longer. I can feel it."

Because Tahnee comes from a very small reservation surrounded by non-Native communities, and went to school with kids who do not know her or her people's history or concerns, she thinks about how this fits into the environmental crisis, a crisis that may seem new to them but that emerges out of a long history to her. She speaks about a larger view.

"It's important for the surrounding world to understand our trauma because it's really a shared trauma. It doesn't affect just the people who experienced it, but it also affects the land that we all share and the society that we're all a part of. While we are Indigenous people, Native American, we are still also American people. And so healing with us is also healing the country and the divide that is still there today. It's really to the benefit of all people in this country to fully address that trauma."

She urges a specific approach, a pragmatic view of how to approach the problems. "As for how to accomplish it? There is a lot to see there. It's not only one solution. I wish it were just an all-tied-up-with-a-bow easy thing. But it's not, and it's going to be different for every tribe. That's important to remember, that there's not just some pan-tribal experience where we all suffered this one thing and there's only one solution to that problem. Different tribes all have different histories. So tackling it one by one and really addressing what you need in that community is important. As for people trying to tackle it as a whole with just one apology for all Native people, all people who experienced slavery or oppression throughout history, it's not a one size fits all. I think it's definitely a difficult task to complete, and that difficulty is what's preventing it, because it's not easy.

"It's not simple, but it needs to be done." With these words, she comes back again to her most fundamental view and sense of urgency. "And it's for the benefit of everyone."

With this admonition, Tahnee leans back and smiles. Her large and handsome dog walks into the picture and comes close to her for a comforting hand on his back, then walks back out of the room satisfied. Tahnee laughs, and we move into talking about her long-term goals in light of everything going on, and her experience.

Tahnee sees her education as a tool for implementing her vision of helping others, both within and outside her immediate community, so has joined a program at the university toward that end, and she expands upon its goals and how they fit with her own.

"The Renaissance Scholar program is set up for people who were previously incarcerated, were in the foster care system, were wards of the courts, or orphans. They provide the needed extra assistance, because people in that category don't have parents or people who can help them. They don't have the support system that most people going into college do, so it provides everything. It's hard going into a university and not knowing what you're doing. I literally had no idea about anything I was supposed to do. I just knew that I signed up and I got in and I didn't know what I was in for or how to do it. So, programs like this help students get on the right foot. They give that little assistance that students need to truly be at the same level that most people already start out in."

She describes the different faces of the people on the program website, pointing to the variety, then coming back to her points about relations. "It's made up of different people and they look unrelated, but you can never really see just by looking at someone, what their story is or where they come from." She focuses on their common need.

Building more on her reasons for being in school, she maintains that this fits into her tribal sense of community, again looking at her personal story within the surrounding world, as she goes into her specific goals.

"I'm majoring in psychology, and when people ask why I wanted to do psychology, I always say I was drawn to it because it was interesting. The first psych class I took didn't feel like a class. It was fun. And that's important, but now I realize the truth of why I was drawn to psychology. I wanted to understand what is going on in me, inside myself. I was still searching for those tools of healing, so it started out as a desire for finding tools for myself.

"But once I get those tools, I'm not going to keep them to myself. I have to share them. I have to share the knowledge with my community because our main struggles right now are mental health. I know that's true for a lot of tribes out there and for a lot of Indigenous peoples out there, a huge thing that we still battle with."

It's important to her that she describes the needs of Native people both within and outside of tribal communities, because far more Native people live on the outside, and she includes their needs as well. She also talks about her community's limitations and what this reflects.

"My tribe is well-established, but we don't have any kind of in-house pro-grams for our lineal members. We have a huge prevalence of drugs and addiction, and the way of dealing with it is just booting them from the rez and having somebody else deal with it.

"I would like to have something in-house, something available to my people right there, something that we do, so we don't let an outside party deal with our people's kind of healing. We need people who are experts in intergenerational trauma in order to really address that issue."

She stops to reflect on a recent incident that hit her hard, underscoring the need she wants to convey. "Also, people really only ask for help once it gets really, really bad. Last year we had a suicide on the reservation, and she called for help. She called the police department and they didn't do anything about it. And then a week later she killed herself."

Tahnee's frustration shows on her face, a heavy sadness at this unnec-essary loss, and she points to the underlying issues within her community, one of many that has been historically excluded from having its health care needs addressed, needs which often remain little understood.

"People don't ask for help until it's too late. It's like when you get a cut on your leg or a wound and you don't go to the doctor until it's infected and you're on your deathbed. You need to go in before that, and you need to know that preventative methods are there. I want to establish this in my community, a culture where people feel comfortable asking for help when they need it, but also a culture where the process of healing is ongoing, all the time, with the knowledge that it's always for all of us. And, I would like to start healing ceremonies for my community."

Without hesitation, she repeats that her small tribe is representative of the whole. "What goes on in my community is a reflection of the world. We're in pain right now. My tribe is, and the world is, and when you look at everything that's going on right now, it feels too big to fix. It feels un-achievable, unattainable, and it's just too much for one person to even com-prehend how to handle."

This feeling of powerlessness has consequences to her. "So we shy away from it, push it away and try to ignore everything that's going on." She shakes her head slightly, then offers a solution, always coming back to her sense of pragmatism.

"When you look at it on a smaller scale, like looking at my reservation, I know there are things that I can do there to make a change. There are things that my community needs, and things that I can do to help. When I tackle it on that scale, it doesn't seem too much to deal with. So look at your community and try addressing it first, because if everybody does that, that's how we start to heal the world."

She addresses a prevailing sadness, again offering tools through listening to it. "Emotional and intellectual climate is tied deeply to physical changes in the environment. People really feel that truth now in California, with all the fires going on. It's devastating and sad, and when you see the images of trees in the rain forest being cut down and animals going extinct, it really hurts. We should listen to that, and realize that those are our relatives, and we're losing them.

"And why do we feel this way? It's because we're connected to the Earth and everything around us. We feel the responsibility that comes from that connection, and it's affecting our emotions. And yes, it's all very deeply tied together."

Tahnee talks further about responsibility, sits in silence for a minute, then leans forward to detail a list of "small" things people can do, from diet to consumption to gardening to driving less, all the small but important contributions many people know about. She maintains that the way we spend our money has a powerful influence, one tool in a vast array of tools, and she is not content with any sense of feeling powerless.

"When you look at those little things that you do, and you're just one little person, it feels insignificant in comparison to what a lot of the big corporations are doing to the planet, to the damage that's being inflicted. It feels like our small contribution is not enough in the face of what they're doing.

"Again, a lot of people just give up then, but by saying that it's all the corporations' fault and by saying that there's nothing we can do about it, that it's up to more powerful people, that deflects the responsibility, when it is always our responsibility to take care of the planet, each and every one of us. It's our responsibility.

"We are a lot more powerful than we make ourselves to be because we are literally creating the next generation of people. A lot of it is a change in mind-set, so the most powerful hold against climate change is getting back to the community roots, getting back to really caring for one another again, and as a result, caring for the planet. We have the power to do that. We have the power to instill values in our next generation that will change the world."

Everything Tahnee says, including the telling of her own story, is one voice among many, in a very long sense of time. She says that she is a small part of something very large and ongoing that helps her see, and this vision rises out of her connection to the land and people she comes from. She says, "I am nobody," and an elder reminds me that when Black Elk said this very same thing, he meant that we are "nothing," while the Medicine is everything. It's the Medicine, and Tahnee knows this. Her purpose and deepest knowing come together in her final words, rising out of her life, spoken with a firm and comforting smile.

"We're not powerless in this fight."

14

Melissa K. Nelson, PhD (*Anishinaabe/ Métis [Turtle Mountain Chippewa]*)

Dispelling Delusion with Alchemy

COMPOSED BY DAHR JAMAIL

No fear, open eyes, open heart. This whole country is carrying this collective grief, and it's about letting that grief be felt. That is part of the wound that Eduardo Duran calls the soul wound that is part of this nation. And until that soul wound is addressed, frankly, honestly, and openly, with studying genocide and the truth of what happened here, it's going to be a trauma that is going to twist up our hearts and minds. Only in the sunlight of understanding can this wound begin to heal.

—Melissa K. Nelson, PhD

Melissa Nelson describes herself as "an ecologist, writer, editor, media maker, and Indigenous scholar-activist." A professor of Indigenous sustainability at Arizona State University, Melissa has been leading the Cultural Conservancy, a Native-led nonprofit focused on cultural preservation, Native land stewardship, and the revitalization of Indigenous knowledge and oral traditions through media and education. She is particularly interested in Indigenous food sovereignty, and is the co-editor and contributor to *Traditional Ecological Knowledge: Learning from Indigenous Practices for Environmental Sustainability* (2018), as well as a co-editor of *What Kind of*

Ancestors Do You Want to Be? (2021). She is also the editor and a contributor to *Original Instructions: Indigenous Teachings for a Sustainable Future* (2008).

We began our conversation by asking Melissa if she would speak toward the general questions of how humans arrived at this point of converging crises, how might we move forward in the best way possible, what stands in the way of our moving forward, and how might we comport ourselves. She began by pointing out that many stories and teachings have come to her, as she has been dedicated to being a good listener and a good learner.

"I get that from both my Native tradition and my Norwegian father, who was very curious about life and always a deep listener and learner of stories. So I'm merely just sharing the stories that I've heard, because I've had these questions from a very young age too, about 'How did we get here? This doesn't seem right. Something is a little off in our direction. Where did humanity take a wrong turn?'"

She believes there was some mistake in the way humans started perceiving a separation between "you and me."

"We stopped seeing that the air that I exhale, you inhale, and that what you exhale, I inhale. That's maybe a lesson of COVID-19 right now, like this virus is somehow showing us that your air is my air, and that we share this vital substance. And it's out of balance right now. It's potentially harmful, from COVID to air pollution."

Melissa thinks that somewhere along the line, the human ego began to seek differences between us, which created competition and a better-than/worse-than thinking, a comparison mentality, a one-upmanship.

"I don't believe in the interpretation of 'Darwinian evolution' that life is about competition. We know from our stories and our traditions that life is about cooperation. It's about kinship. Yet the narrative of competition and individualism and one-upmanship became valorized in European sciences and the Renaissance, and the separate 'knower' that has destructively colonized the world and colonized our mind with fragmentation to think that we are individuals, we're competitive, it's either kill or be killed."

This has, as she sees it, led to a "survival of the fittest" mentality, and the idea that the accumulation of external objects is somehow important, more important in fact than giving away and sharing.

"That was a poisonous worldview that then spread like wildfire through imperialism and colonization. And we know that nobody is immune to it, as it infects every color, creed, and religion, and yet many of our Indigenous ancestors were immune to it."

Steering the conversation toward that pivotal point, Melissa remembered how her ancestors pointed out how ridiculous the notion of separation and all that went with it was.

"We have to live in a way where we give things away, we don't accumulate, because accumulation becomes toxic," she said while she held her hands out to underscore how clear of an assumption that was. "And that's what I think we're seeing right now . . . this gross accumulation at the cost of the environment, of the land, of forests, of species, of minerals in the Earth that are doing their job. The blood of Mother Earth is being mined and turned into plastics and products for human entertainment. It's just completely backward. So we went wrong with this idea of accumulation, of personal gain, of hoarding, of competition, the greed of the acquisitive mind."

Melissa believes this needs to be rebalanced.

"If we look at what people really love, we love our families. We love our partners. We love our friends. We love to share, give away, and benefit others, like those sayings that remind you that your best work is the work that you don't get paid for. Most feel that service, helping others: that's our *real* work, right? That's what many say. I'm very fortunate that with nonprofits and universities I do get paid for what I love, but many people often say, 'You have a day job and then your real work is volunteering or offering your services to help others, elders, children, or animals, or various causes.'"

Given how most people know that giving, not taking, is what life is really about, Melissa asked, "How do we move and transform that in the world, and especially in America, with this current outgoing president [Trump], windigo-in-chief, as Robin Wall Kimmerer called him."

Melissa referred to windigo, or wétiko, the term Native American scholar Jack Forbes used for what he described like this: "Brutality knows no boundaries. Greed knows no limits. Perversion knows no borders. . . . These characteristics all push towards an extreme, always moving forward once the initial infection sets in. . . . This is the disease of the consuming of other creatures' lives and possessions. I call it cannibalism."

"It's just like a poison to the mind," Melissa said. "And that's when I do get scared. This poison of the mind spreads so far and so deep, and it's not going away. Yet we need to find an antidote to that poison."

Melissa believes deeply that the antidote is in Indigenous values, Indigenous wisdom, and Indigenous sustainable lifeways, but that is going to require a massive transformation.

"I think it's going to be 'breakdown to breakthrough.' I don't think it's incremental change. Our version of democracy that we've seen in America right now is hanging on by a thread because its foundations were faulty. They only adopted a few things from the whole."

The "whole" Melissa referenced was the Haudenosaunee/Iroquois Confederacy in northeastern Turtle Island whose great law of peace originally shaped U.S. democracy.

"They [founders of U.S. colonial settler democracy] forgot the biggest thing," Melissa continued with a smile. "Women! Mothers! We have the ability to revoke, to impeach, to take out those chiefs. And children were also part of those decisions, as they were seeing further. So the Founding Fathers did not get it all right. They forgot the women. They forgot the children, the value of intergenerational thinking. They forgot a few things!"

She sees what happened at that point in history as a co-optation of a Native tradition.

"Putting the eagle up there with the arrows," Melissa said, speaking of the presidential seal, "it was a co-optation and it was another symbol of conquest. They tried to pretend it was honoring Native traditions, but it was a stamping out, and almost like a mascot the way the eagle and the arrows were used by the Founding Fathers for the United States of America."

While Melissa is grateful for the attempt at democracy in the United States and feels it was a step in the right direction, she believes everything is going to have to be transformed at a fundamental level. She does not know exactly what that is, but thinks it certainly entails going back to local traditions, honoring the Indigenous peoples of wherever you live, serving them, volunteering with them, and being a good ally.

"Helping with land rights, land reclamation, food sovereignty. We have a lot of solutions, like council-style consensus decision making, more women in power and in leadership positions, more youth in leadership positions. I think those are all ways we can help to restore and transform."

When I read Jack Forbes's book *Columbus and Other Cannibals*, I was finally able to understand how people in power, and those who support them, could carry out policies that are literally killing life on Earth, including their own loved ones and children.

Given that Jack Forbes was one of Melissa's professors at UC Davis when she was attending graduate school there, I asked her if she would discuss wétiko disease more in depth. Reading his book, which she called "just incredible," was one of the reasons she chose to go to the school where he taught.

Melissa spoke fondly of the late professor, sharing how he did not take his lunch in the faculty club, but instead ate with the students at the same time and place each day in order to make himself available.

"It was pretty magical," Melissa said of that time with Forbes, speaking with a warm smile. "He reintroduced me to the idea of loving our grandfathers and foremothers, the wise ancestors of many related nations, the knowledge keepers who share the same Algonquian language."

Melissa related directly to what Forbes described in detail in his book, as her mother remembered the stories from her elders about the cannibal in the woods that would come out to devour everything in sight, along with the direct memories in her family of boarding schools, the role of the Catholic Church, and other atrocities carried out against her people.

Speaking directly of wétiko, Melissa saw it as an archetype in the psyche of Native people.

"We're brilliant enough to turn it into this incredible story that anyone from a child to an elder would understand and learn from and be afraid of. This is an example of some healthy fear of what happens when we let our most base nature take over, which is also referred to as the seven deadly sins. There are many different ones from many different traditions, but greed is a major one that has infected our society. And wétiko represents greed in the form of appetite and hunger, and being insatiable for external consumption to try to fill an internal emptiness. When you have a spiritual bankruptcy you try to fill it with these external things . . . but it will never be filled by things. Never."

Melissa paused for a moment in thought, then continued.

"Wétiko is filled by this insatiable hunger that consumes everything it sees. It knows no end and has no limit. There are many stories about how to put it down, and there is usually a sacrifice involved because wétiko is so powerful that someone or something generally has to be sacrificed to defeat that power. That is very telling and curious, given how in the world, and especially in America today, there have been so many sacrificed already. Children, forests, immigrants, Indians, you name it. Like we haven't sacrificed enough? And that is where that profound frustration comes in around how much has already been sacrificed, yet wétiko just keeps building up power, corruption, and destruction."

She then expressed how we need to go deeper into the stories of how to defeat wétiko.

"I look to the teachings of the Anishinaabe and the Haudenosaunee who've also been great teachers of mine, John Mohawk, Katsi Cook, and Oren Lyons, and others. There is the story of the great law of peace. And of Tadodaho, the evil wizard who was wreaking havoc on the five nations, causing blood feuds and destruction. The peacemaker came and transformed Tadodaho, not by destroying him, but by elevating him. It's an interesting kind of aikido move. Many of our Native stories talk about how you don't destroy your enemy, you elevate them, you transform them. And by elevating them, they understand that their ways have been evil and destructive and they have a change of heart and a shift occurs, so that they then become responsible, giving beings, rather than taking beings."

Melissa wonders how this could be accomplished, given that wétiko thrives on battle and fighting. She acknowledged Stan as a war-era veteran, and myself as having witnessed war firsthand in the Middle East, and used us as examples by way of what we've both endured. She was reflective of the need to find a peaceful form of transformation for wétiko.

"Battling just isn't going to work, and that's confusing because the warrior spirit is also very strong." Speaking directly to Stan, she added, "You know firsthand, Stan, of the long tradition of Native warriors. So where do you meet this spirit from both a warrior perspective and the peaceful transformation?"

Melissa thought it might be some form of triangulation that involves both approaches, given the current "evil that is unleashed" in our country, and across the world.

But she also saw wétiko teaching us about this in ourselves, which needs to be looked at.

"That's why we have our original instructions, that's why we have our ceremonies, that's why we have our teachings that have to be practiced every day because they are our immunization. They are the antidote to the wétiko spirit that's in all of us. Because we all have the seeds of greed and selfishness and competition. But in our original Anishinaabe instructions, we have our seven grandmother teachings that are about love, bravery, respect, humility, truth, wisdom, and generosity."

She believes these things, and a broad solidarity, must be practiced. They cannot just be empty words. Melissa expressed that we can look to our allies in many different circles for the solution. Community keeps us accountable. Using herself as an example, while she is critical of mainstream religions for many solid reasons, particularly because of what they've done to her ancestors, she has done work with the interfaith community.

"And with the boarding schools and other mission efforts, I think that many of the world religions too are starting to grapple with the destruction that they've wreaked on people of color in particular, and on the planet, and are trying to transform to some more holistic and healing modalities with each other."

Melissa said she was encouraged by this, albeit in a cautiously optimistic way, before adding that she believes Indigenous people should be leading this work at local levels, given their strong traditions of peacemaking and restorative justice.

We asked Melissa if she believed a strong dissection of the genocide of the Indigenous people of Turtle Island could be a part of this elevation of the human spirit, given that all of us who live here are, quite literally, children of genocide, regardless of what side your bloodline falls on.

Since Melissa has been teaching American Indian studies for decades, she has always believed in the need to address the genocide directly and immediately.

"No fear, open eyes, open heart. This whole country is carrying this collective grief, and it's about letting that grief be felt. That is part of the wound that Eduardo Duran calls the soul wound that is part of this nation. And until that soul wound is addressed, frankly, honestly, and openly, with studying genocide and the truth of what happened here, it's going to be a trauma that is going to twist up our hearts and minds. Only in the sunlight of understanding can this wound begin to heal."

Melissa sees wétiko filling the void that occurs when the trauma remains unresolved and unhealed, and believes this is a global problem, not limited only to the United States, and sees it as a critically important issue.

One of the reasons she thinks the genocide of Indigenous peoples is not being studied is the shame and guilt of the descendants of settlers, which is also part of the wound. This wound is what she sees as driving the White supremacy that has been plaguing the United States since its inception.

"It's unaddressed, it's valorized rather than being healed. So absolutely, genocide studies need to occur locally, how it happened wherever you live because that hits close to home, and it's tied into things like place names. Why is Mount Diablo called that? Because Ohlone people were having ceremony there and the Spanish saw them and killed them because they were 'devilish,' because they were having ceremonies with fire and furs and skins that the Spanish didn't understand."

This is just one example underscoring the importance of studying genocide locally, given how it directly ties into place names so often.

"We have Sutter Hospital in California. [John Sutter enslaved, abused, and slaughtered Native Americans.] He was just horrific. People don't understand how we have the names of streets and hospitals and mountains in places, and that these are ongoing forms of genocide of Native peoples. Until that's expunged and purged, or at least until the truth comes out about how it's linked to genocide and erasure of Native peoples, and they go into their own shame and guilt and infection with conquest consciousness, people are going to continue making grave mistakes. Without truth and justice these horrific relationships will continue."

Melissa then asked herself the question again of how all of this can be framed as an elevation of the human spirit.

"Because we're also still here," she said, while looking directly at Stan. "You and I, we are still here. Our ancestors, despite all odds, are still here. Even though our families went through relocations and brutal displacements, many survived, and Kyle Whyte talks about this in terms of how Native people have already experienced climate change.

"When you take your people, the Apache, and Geronimo, out of the Sonoran Desert homelands and put them in the Everglades, in the swamps of Florida, that is climate change. When you take the Potawatomie out of the lower Great Lakes and send them to the plains of Oklahoma, that is climate change. So our people have been through these radical shifts of life. First you're there. Now you're over here. Different people, different mind-set, different landscape, different climate. And yet many survived. We're quick learners. I think one of the most important aspects of Indigenous peoples that I've seen here and all around the world that I've been privileged to visit, is the ability to learn and adapt, to keep a very open mind, which means a humble mind, because we're always learning and adapting."

Melissa paused there, sitting with those thoughts for a moment, before continuing.

"Great Spirit, Great Mystery that we prayed to at the beginning of this, is always there, always with us. We know nothing. We know a smidge, right? So it's an elevation of our spirits so that we can see how our people survived and adapted. And we can also deconstruct the colonizer worldview and the people who inflicted genocide, and try to understand and transform that mind-set."

Melissa knew we were in intense conversational territory, and pointed out how both Stan and I have to work constantly in that realm, given the PTSD from witnessing the war machine firsthand, and through that work continue to strive to become healed, elevated, and transformed.

"We've seen that, over and over again, it can be healed and transformed by learning the truth and reconciling it within ourselves and repairing it. There are beautiful traditions, right? The Lakota have the wiping of the tears ceremonies, the Hawaiians have the Ho'oponopono forgiveness ceremonies. That's part of Indigenous traditions too. We *have* these healing ceremonies for our people."

She spoke of the need to elevate these, and cited Canada as an example of having attempted to have a reckoning with truth and reconciliation around the residential schools there. While there are mixed opinions on how successful that attempt has been, Melissa pointed out how at least now those schools, and the atrocities that came with them, are part of the conversation in Canada.

"People know that Indian children were stolen from their homes and abused and often perished in these horrible residential schools. You can talk about it. But here in America, you bring up the boarding schools, and people think of rich White kids going to prep schools for Harvard or something. And that ain't what we are talking about here. It's not Harry Potter."

We all laughed at her gallows humor.

Melissa wonders why so many people are afraid to look at the truth of what was done to Native Americans, because in so doing they are living a lie, and a delusion. She sees the job of elders, educators, and teachers as dispelling these delusions.

"That's where our warrior spirits come out. Dispelling delusion. I love that about the Buddhists, the Tibetans . . . they have the swords, and they cut the heads off of illusion. We really need to do that again."

Melissa had earlier used the term "miss-take," which intrigued me, so I asked her to talk more about what she meant by that.

"When we talk about mistakes, most of us immediately go into denial and shame. But just breaking that word up into its two key components, it's a 'miss-take,' and it's deeply related to the word 'respect,' which is to re-spect, look again, as in spectacles; it means we are really going to look. Miss-take is when we don't pay attention, when we just let our conditioned thoughts determine something like racism. 'Oh, they're this? Oh, they're that.' We have all these conditioned thoughts so that we don't meet people freshly. We don't meet the Earth freshly. We have all these preconceptions and assumptions. It percolates when we see something. So by breaking the word down, it's a miss-take: we don't look or see clearly."

Melissa preferred to phrase it this different way as a reframing that re-moves shame from mistakes we've made, by noting these occurred simply because we did not see something correctly. She learned this teaching from the late physicist and philosopher David Bohm.

"It slows me down," she added. "It puts me in that humble space of a learning spirit to look freshly, to remember to look at my partner of thirty-three years in a fresh way each morning, and to remember to look at red-wood trees and really take them in."

To her, that kind of perceiving and opening ourselves means that we have to be more vulnerable when we respect something because we're looking deeply and opening our senses much more.

"It's receiving the spirit of that tree or of that person and that energy ho-listically in our heart and our body and ourselves, not just with this idea I have in my head. So 'miss-take' really helps me understand perception, and understand respect, to try to get a clear perception of what's going on rather than this idea that I have in my head.

"All those misconceptions and assumptions have to be peeled away so we can see the world fresh again," Melissa summarized. She spoke to how our hearts will break when we do this, as all three of us knew, again, from our work with PTSD. "But there's so much resistance against understanding the foundation that it's all built on. Until we go back into the soil and renourish and heal those roots, we will keep making these mistakes."

At this point Melissa felt the need to address the despondency and powerlessness that has become so prevalent amidst the converging crises. So many people feel that there is nothing to be done about what is happening now, or even that there's nothing to be done about the genocide because it was so long ago. Speaking from her own experience, again, Melissa believes that elevating people is the best way to activate them.

"That's really been a lot of my work as part of healing myself, from the traumas that I inherited, and that we all inherit being American. And certainly being Native American."

Melissa sees it as important that, collectively, we get to solutions that lead us to feeling we are responsible, have a role, and that there is work that can be done where we can make a difference.

"That's what gives people hope. What's so sad to me is seeing young people who feel like they have no hope; there's nothing they can do that can change anything, so why even try? So just get blotto on whatever form of self-medication or entertainment or violence because there's nothing to live for. That is so sad to me. Yet everyone has a role in this collective malaise, and there are also profound structural inequities that make things more hopeless for some."

Getting back to the basics is what she was teaching in her classes when we spoke with her. She asked her students what they really needed to live. While many pointed to cars and computers, she reminded them to get even more basic, to remember things like shelter, water, air, and food. While she spoke to a certain lack created by consumer culture, she also spoke to the good changes she sees in many youth.

"We have so lost touch with getting back to basics and what it takes for human beings to live on this planet, a reallocation of priorities is necessary. I see incredible inspiration with young people around food and water, and the food sovereignty movement, the sustainable agriculture movement, seed saving, seed rematriation, and returning to our seeds and returning to ancestral lands as forms of Indigenous resurgence."

Melissa believes it is imperative for us to dig our collective hands into the soil to reestablish that primal connection that is so vital to our mental and physical health. She looks very forward to the post–COVID-19 world when she can again gather in groups, and resume taking students to places like Indian Canyon, to Alcatraz, and other important Indigenous sites to teach the true history from the Native perspective.

Melissa has watched young Native students' eyes light up when they learn the true history of what happened on the lands they live on. When they learn of destroyed sacred sites or massacres that happened in their and others' ancestral lands, they are directly impacted and changed by this knowledge.

"Some of that traumatic colonial truth is best digested when you have a visceral experience of it. Reading books is good, and hearing stories about it from people who have been affected by it is good, but actually going to those places after learning about them and offering some kind of healing, and then receiving some kind of healing, is so important."

Melissa has been inspired by the Native youth she has seen already involved in things like the Water Protectors movement that became elevated by Standing Rock. This is because it had already been such an important part of their culture, given how women had always been water keepers, and cited examples like the Ohlone of the Bay Area who are linking up watersheds, including the Sacramento River watershed, from its headwaters, as well as the Winnemem Wintu of Mount Shasta linking to the Ohlone at San Francisco Bay, in order to bring back the salmon.

"That is an act of healing," Melissa explained as tears welled up in her eyes. "You learn about the trauma of the land, the genocide that happened there, the villages that were there, and you learn about the contemporary

Native people who are there, who are struggling but living and bringing back their basket-weaving plants and their salmon and giving something to that river."

She began to speak directly to the river, her voice rising, filled and wavering with love, as the tears kept coming. "River, we have not forgotten you. We love you. We're trying to heal you. We're singing to you, please meet us halfway, come back. We have not forgotten you."

All of us were filled with emotion, quiet with the energy Melissa had brought into our meeting, our hearts summoned to the rivers by her offering. We sat in that feeling, breathing.

"We call it many things," Melissa then continued, speaking gently yet matter-of-factly. "In academia we often call it ecocultural restoration because you gotta bring together things that English language separates, right? Process remembering. So it's a lot of translation too." She paused and looked at Stan's side of her computer screen and asked, "Right, Stan?" then continued without missing a beat as he nodded. "Translation of our traditional values into a language that people kind of go, 'Oh, cool. That's intriguing.' Right? Restoring the land and the people, ecocultural, biocultural whoop! Great!"

She laughed hard, as did all three of us.

"Gregg [Castro] said now that they can give us an acronym, TEK, they can now hear from us," she said while holding up her arms as if in victory, as we all continued to laugh.

Stan, laughing really hard, concurred. "Yes! Exactly! We've succeeded! And it's even all in CAPS!"

Melissa, still laughing, continued: "So vindicated! Oh my goodness. Yeah, there is a lot of work that needs to be done. So when people say there's nothing I can do, I'm just thinking, 'Where are you?' And yet I hate giving people these laundry lists of things that they can do. But honestly, some people like that. You know, I try to want to encourage the self-determination of people to find their own solutions, find their own ways of contributing positively to the healing of the planet and our peoples. But some people really like those to-do lists and sincerely want guidance."

From there, she mentioned that she hadn't spoken about decolonizing conquest consciousness, but had begun to discuss the manner of how we might comport ourselves, so I asked her if she would continue to speak into that.

"There's so much hope and so much wisdom available to us, if we ask for it in the right way. There are so many good case studies and lessons of what not to do, but what *to* do? What about how you *know* what to do? And it always starts with ourselves, right? A daily practice within ourselves, whatever that is . . . supporting people in having some kind of daily practice. It could be so simple, a little offering of tobacco. It could be a prayer, an intention, remembering those original instructions."

Melissa has consistently told people that sustainability starts with breakfast, by which she means by what you put in your mouth. Where did the food come from? Because every choice we make, every day, has an impact on the planet, as so many people say, and yet so many continue to fail to make better choices. Again, access and inequity are part of this issue. Some people are forced to live in places with little food choices, for example.

"There's just not as much thought put into how we're all interconnected and how 'my' actions affect people in my neighborhood and around the world. So education is certainly a key part of that, both formal and informal, and on the ground land-based education, which is really growing now in Canada. The First Nations people are getting people out of the classroom, and we need to do more land-based education like that in the United States, to give people a sense of how to get back to basics and what it means to carry water and cut firewood and grow your own food and process your own food. Those are lessons that most people today take completely for granted and have no idea where their food comes from and how to grow it, how to tend to it, or to harvest it."

Melissa used food to link into the climate crisis.

"I find great hope in the food sovereignty movement and people's love of food because we've all got to eat. We all love food. We eat the environment every day. We eat the land every day. It's our most intimate connection with the land, besides the air, but people don't really see air as a thing because it's invisible. But it's the spirit of these things, and that is vital. So to give gratitude and to honor those basic elements again is part of what keeps me grounded and keeps me hopeful that we can get through this."

Melissa paused, collected her thoughts, then continued to speak, in a matter-of-fact tone.

"Our people have gotten through it before. Not only the climate change example that we talked about, but with epidemics. Maybe not the pandemic scale we're at right now, but our people went through epidemics of extreme proportions. Dying left and right. Whole families, whole tribes, full clans, smallpox, and all of that. So in our family traditions, we also have stories of surviving epidemics and plagues. That gives us great resilience, and the ability to really honor those resilient traditions by being as adaptable and fluid and strong as possible."

Melissa had briefly cited the concept of "Survivance" from the book by that title written by Anishinaabe author Gerald Vizenor. The concept highlights a Native active sense of presence, which supersedes a colonial history of Native erasure and oblivion.

"In Vizenor's concept of survivance, one of the key factors is humor, and the role of humor as a survival strategy, and how important it is to keep that sense of humor, despite the seriousness and gravity of the situation right now," she explained.

"I find that always so helpful. Another concept that he uses is trickster consciousness, which is another way to get out of the habit of thought about projection and assumptions and to bust out of those patterns of thought that are often self-destructive and harmful to others, because trickster is always waiting in the wings: Raven or Coyote, reminding us things may not be as they seem, and that we need to maintain humility and a really deep openness to what's happening so that those learning spirits can keep guiding us from our gut and from our heart, and from our other sources of knowing, not just from the mind knowing."

Melissa thinks that the transformation being asked of us to make it through these times is almost akin to "becoming new human beings. I don't know if we're going to evolve physiologically, or if we are just going to evolve in our consciousness, but it seems there needs to be a whole new transformation of the human being."

She talked about a prophecy that could be upon us now.

"We're not there yet, but in the Eighth Fire Prophecy, it is a new kind of person that will emerge if we're able to come together and light the Eighth Fire. That will require not just the Native peoples' solidarity. It means we heal with the colonizer and with those who brought so much destruction."

While hopeful, Melissa said she did not know if that was going to be possible or not.

When Melissa had spoken of and to the river, another world welled up within me, and what was strongly shared between the three of us felt alchemical to me.

I wondered how, when we really look into what has been done to Indigenous people everywhere, and all the atrocities done to the planet, this changes us inside. I asked her if she felt a sort of spiritual alchemical shift is what is required of us now.

"Thank you for the word 'alchemical'; that's such a beautiful word. This will get more personal now. I, and we, need, collectively, as you both have done with your own trauma as an ongoing process and a daily practice, to not be afraid of our own trauma and our own shadow. To feel what's happened to the planet, to our people, to anybody."

Melissa believes we need tools for this, and modalities of healing that give us a handhold, because without that, doing so would simply be overwhelming.

"It's too scary. So we have to break down to break through, yet the breaking down is so frightening, we think we're going to disappear."

Melissa mentioned the movie *Wiping the Tears of Seven Generations*, and believes we need to "wipe the tears of the seven generations since Columbus arrived in North America," and this needs to be done regularly. She cited the ancient Hawaiian practice for forgiveness and reconciliation, Ho'oponopono, as a perfect example of doing so.

She sees this as starting at a personal level, then to be done locally, then at larger levels to promote the alchemical change.

"And it is alchemical, literally. The chemicals in our bodies change when we are in that 'fight or flight,' or freeze, and in trauma and fear, and all the cortisol that goes with that. Then we break through, and the dopamine and serotonin and other chemicals flow through."

Melissa pointed out how, whether it be via EMDR work, dream therapy, plant medicines, meditation, or other methods, our patterns are broken down and fresh insights arrive to alchemically alter us in ways that allow us to see more freshly.

"Like meeting that redwood tree like a newborn would, with all that wonder and beauty. Sometimes it may be grief, if it's a stump or in a polluted area. So what is required is a willingness, a courageousness to be truly open. To feel all of that is key. It's central, and I think we all need to do more of it; I certainly do."

Melissa did not mince words, knowing that this is a scary process for anyone to go through.

"I sometimes fear and wonder if the tears won't stop. We think that the tears won't stop and yet, they do. They come and they go, they're like rainstorms blowing through. It's just part of the beauty and power of the Earth and our own internal climates."

Growing pensive at this thought, Melissa wondered aloud about the synergy between global climate and our internal human climates.

"I'm thinking about how climate change is also about our own internal climates that have become out of balance; it's kind of the mirror of human consciousness being like a climate in chaos. They are so interrelated."

From this perspective, the climate crisis is simply a mirror to what has been ongoing within humans in terms of the imbalance, flooding, droughts, wildfires, starvation, and migrations. The three of us were deeply moved by the idea. Stan reflected aloud how Ilarion Merculieff spoke about our upside down society, how the mind is telling the heart what to feel, when the heart should be telling the mind what to think. He reminded us of the writer, poet, storyteller, and culture bearer Darryl Babe Wilson's (Achumawi and Atsugewi) story of the Mis Misa spirit force within Akoo'yet (Mt. Shasta) that sings to keep the universe in balance, but stops singing if people cease to listen, at which point everything goes awry.

In this way, Melissa brought another world into our talk when she spoke directly to the river, bringing us all deeply into our hearts, and he thanked her for doing so, and stated how this is the place where we need to be in our daily lives.

"Absolutely," Melissa responded immediately. "That place of vulnerability. For various reasons, so many of us are taught to not be vulnerable, and we have good reason not to be vulnerable, with all the armor that we build up from life experiences, trauma, war, disease, famine, or whatever it may be."

She paused for thought, then continued.

"So, how to find solace again in that vulnerability? And to see it as a well of healing rather than a well of grief?"

I sensed how Melissa and Stan were articulating the necessary components of our collective healing, and thus that of the planet.

"How do we really take our freedom to the max," Stan asked. "It's like you started out saying, which to me is profoundly beautiful, that everything you're saying is not you. It's everything you've been given. That it is everything you've been taught. And that's the spirit of the book for us."

Melissa immediately picked up the thought and feeling.

"It's like that flow of that river again," Melissa said emphatically. "We are merely just passing through, and if we are fortunate, this perennial wisdom is passing through us, and maybe even some of it will have our own little unique signature on it."

She then spoke to our shared experience of talking together, analogous to what could be achieved more broadly by us all.

"To be in that river and flow of meaning and wisdom and teachings that we've been in here," she said softly, "I'm just humbled and blessed to have a little slice of this on our journey. That's what you'll be offering with this whole collection. It'll be beautiful. It'll be a waterfall. It'll be a gushing water well."

15

Kanyon Sayers-Roods
(*Mutsun Ohlone/Chumash*)
Cultural Competency

COMPOSED BY STAN RUSHWORTH

We are going to need a focus on honoring truth in history, and we have to have
these civil conversations that both validate the hurt and also help strengthen
community members' ability to navigate uncomfortable conversations.

—Kanyon Sayers-Roods

Kanyon Sayers-Roods is a spokesperson of Indian Canyon, a community
south of Hollister, California, that provides a ceremonial gathering place
for many local Native people, as well as international visitors. During California's devastating mission period, escapees found their way to the canyon to save their lives, and today children from nearby schools go to learn
about its history and the local flora and fauna. They stand in a large circle
and pass the abalone shell with burning sage to one another. Fifth-graders
listen to elder Ann-Marie Sayers tell the history of the canyon, and as
they do so, falsehoods taught to them in fourth grade at the missions
(California schools' curriculum) drift away from their young minds and
embrace another reality born in the people before them. They listen to
Kanyon sing the grandmother-honoring song under the oak trees and
experience an older way of seeing the world around them. I remember

my young son listening, standing in that circle, gifting Ann-Marie tobacco, and her broad, warm and welcoming smile. Kanyon sings that same beautiful and powerful song all over the San Francisco Bay Area, carrying the message of Indian Canyon to great numbers of non-Indigenous peoples, through her work in cultural consulting, schools, community coalitions, land acknowledgments, and event blessings. She has been singing it since she was a child.

"I identify as a California Mutsun Ohlone, two spirit educator, activist, artist, and community member. I am the daughter of tribal chairwoman Ann-Marie Sayers, and I was raised in Indian Canyon, the only federally recognized 'Indian Country' [Kanyon gestures with air quotes] between Sonoma and Santa Barbara on central coastal California.

"That being said, I recognize my privilege, how I was raised in an environment where cultural competency and cultural sensitivity was a norm, as well as recognizing the cultural and traditional protocol of Indigenous community members who acknowledge where we come from and where we are, and how we came to be here."

I think of what privilege means to her, that it's something far deeper than money or power in the conventional sense, but more about how we see and operate. The words roll out of her, and I lean back to listen.

"I've been able to witness community members come respectfully, seeking a consensual interaction with the Indigenous community of the land that we are on. As I grew up, I recognized that this occurrence is not very common out in mainstream society, at least in the Americas.

"I recognize how much of a privilege it is, yet it was hard to see how lucky I was as a kid, and I would complain about being deprived of modern amenities such as pressured water, or electricity that lasted through the night, or all the toys and goodies that I got to see my peers enjoy in elementary, middle, and high school. However, as an adult, I recognize that my lived experience has given me many amazing opportunities."

A key opportunity Kanyon details is her perception of the interwoven connections surrounding her. When her mother talks to the circles of children, and to the many students visiting from local colleges, she emphasizes "right relationship" to the land they're on, and to history, and Kanyon's observations of how we've arrived at the situation of the Earth right now come from this teaching.

"What we in society are going through in regards to climate change, I blame on the behavior that has manifested due to colonization.

"Elders not even thirty years older than me have informed me that their elders would tell them to be quiet about being proud of who they are. Not many know how frustrating it is as a Native youth, going to school and not ever hearing about the history of the land that we're on, not getting a chance to hear accurate depictions of Indigenous peoples, Indigenous pedagogies, Indigenous governing systems and structures and roles of the community, and seeing Native culture being hyper-sexualized and hyper-romanticized and invalidated by outsiders proclaiming that we are savages, that we are deemed closer to animals, and that we have no rights."

She describes her frustration at the sustained attack on a necessary and long-standing way of thought, one being replaced by ideas with a completely different set of goals.

"We're in this mess because of the pride and ego of that outsider coming in. It thinks it's correct, that mentality, that behavior, correct about how it should navigate this environment, because it has not been brought up in a culture that has humility or accountability or reciprocity to sacred living systems. Its behavior has led us to be where we are, a behavior that prioritizes resource extraction, both from the natural environment and from Black and Brown bodies.

"Its decisions are made in ways that may only benefit that entity, making the decisions without accountability to their ripple effect. It is not very humble. It is engaged in a capitalistic, materialistic, resource extractive, and disposable society mentality, a focus point which prioritizes its decisions."

Kanyon speaks with surety, and quickly contrasts specific differences between her immediate community and many of the surrounding efforts toward positive change, and the conceptual borders established.

"It's frustrating, because everything's related to everything. I witness people who want to fight for rights or lives, and I see PETA or animal circles separated from the LGBTQ movement, separated from the teachers union and separated from racial equity circles. But they're all related. If we respected community, if we respected sacred living systems, we would find resolution in how certain decisions are being made. If we were raised in a culture that was accountable, we wouldn't be in this mess.

"However, if we can, we should realign to the best of our ability, and that also means we need to look at the ugly-layered occurrences of truth in history. And to some people that's really, really hard."

Kanyon consistently advocates accountability to layers within history everywhere she goes. To her, this is not something that separates people but that brings them together, if they have the desire to get past the difficulties.

"There are community members who are really for it, and I'm lucky to encounter those beings. In this devastating occurrence of quarantine and the outburst of COVID, I am seeing communities come together, and I'm happy to witness how members are advocating for truth in history. On TikTok, I'm seeing [ethnically] mixed community members voice their truths about how in their environment, in different generations, their education system does not account for their social, cultural, and even economic environment."

In the area where Kanyon lives and works, there are a tremendous number of mixed-race and mixed-culture people, and she reminds me of the impact of silence around this factor in our lives, and what is possible when all the layers are given voice.

"I highlight why it's important to honor truth in history. And I also validate circles learning from each other and attempting to share stories, attempting to acknowledge the teacher within us. This is step one, or a beginning stage of growing and bridging community together.

"Over time as we relearn how we can be together, my hope is that my dream will come true, where more of us will recognize that we have many teachers within our circles. We all have something to share when it comes to youngsters in classrooms who are there to learn in that system. When they see that they can be a teacher, they hold a sense of responsibility in that sharing. We all learn how to interact with other people with respect and reciprocity, for a time."

She pauses, reflects again on the isolation of COVID, then reiterates her happiness at seeing folks coming together online, but at the same time, she is clear about the limitations, especially for someone living in a close community who feels obligations to serve in the best way she can, and who is called to events all over the San Francisco Bay Area.

"I miss going to ceremony. I miss being able to just run up to friends and hug them. I finally learned that I am an introverted extrovert who learned how to synthesize energy from social environments, and I was finally able to acknowledge this as true at the tail end of 2019.

"People would come to me saying, 'You're over at this event, you're over at that event, how do you maintain?' I just said, 'This is how I navigate the world at these times. This is my internal ceremony.' When I'm in social environments, I get a very strong charge, and then boom, COVID hit and ripped away all that source of energy."

She describes a health issue coming onto her, how it all dovetailed, and what she learned from it, listening to what she's been taught.

"It was a combo package, a whammy of both my life and ancestors saying, 'Guess what? You need to learn different ways to be, and you have to slow the heck down. Check in on yourself. You need to make decisions that you know are beneficial, to focus your energy and rein in your efforts because your patterns of behavior are not sustainable.' One deep realization I came to was that when it came to social, spiritual, and emotional burdens or tasks before COVID and my health, I never knew that those took energy. I'd never walked in this world where I had to consider how much energy I needed in order to navigate these interactions."

Listening to her, and thinking about the deep influence of the canyon where she grew up, I think of the elders I know who have to collect strength, to "gather thunder" before going out into the world to help, then gather it again upon returning.

"Now I know this, which is completely foreign to me, and I'm recognizing that there are so many factors in mainstream society we have to consider. I will never know if our ancestors had to navigate and traverse these kinds of issues, like identity crises. I'm 'Native passing'; however, if you try to follow that validation through the stupid colonial construct of blood quantum, they would try to dismiss me. So it's combating identity politics and academic and civility politics too.

"It's about how we navigate these social environments to validate our legitimacy in the eyes of a governing structure who wants to dismiss us or dismiss our narrative because they were never exposed to truth in history."

Kanyon reiterates that it all fits together, that there is no separation between the fracturing effects of personal and social politics and the issue at hand with Earth and all her inhabitants.

"It's about having external factors with deep impact on our lands and our environments, and attempting to reconnect to culture and ceremony and traditional practices and protocols, as well as land management and stewardship methodologies. There are the legalities of not having the right to access our cultural resources, our environments, our means to steward our traditional lands with our traditional ways. I can't just start a fire and manage it because I will have external factors arresting me. I can't go out into nature and harvest basket-weaving resources or pine nuts easily, because I would be deemed invading privately owned property. If I was out on state land, I'd still be in trouble until they forced me to prove all of my genealogy."

She stops for a second, out of patience with the last sore point, then speaks for countless people who struggle with the demand to prove who they are. Few non-Indigenous really consider the damaging effects, especially on youth.

"As Indigenous people, we are the only group that has to validate who we are 'legitimately.' How many times have you ever heard someone say, 'Oh yeah, my mother is Asian and so I'm Asian American.' 'Oh? How much Asian are you?' It's quite frustrating that this is a mentality we navigate in both social and historical contexts, because if we acknowledge how we got here, we would start pointing out what we *should* prioritize in regards to teaching the next generation, but we don't."

She points to the importance of the choices teachers and institutions make, and I'm reminded of the numbing mission curriculum just up the road, and of the difference Indian Canyon has made for thousands of Native and non-Native children. Thankfully, this legacy feeds her.

"There are community members who voice that the statement I shared with them deeply impacted them, and that they are taking active steps in their life to shift how they're navigating. I recognize how lucky I am to be an artist and activist and speaker, getting feedback saying that what I'm doing is of value, that what I'm doing is offering insight. And I'm careful with that. That holds responsibility."

Kanyon returns to her credo of looking at both the past and present, emphasizing that it's something requiring a broad and ongoing range of thought, a tapestry.

"We need to honor *all* of the layers of truth in history, before contact and during, talking about how being an Indigenous person was illegal and not safe. That's why we have the occurrences today where community members say none of our family members had Indigenous ancestry, or the family members who did say we have ancestry, say it in a negative fashion. Why is that, when the community members who might be curious if they have Indigenous lineages in their family start their journey? These layers need to be shared in our schools."

Listening, I reflect that if colonial thought is at the heart of all the destruction, being afraid to explore one's indigeneity is fearing the scope of connection needed, a control mechanism that impedes critical and heartfelt analysis.

"We should focus on our ecological system, our ecosystem, our bioregion, our watershed. We should focus on the layered history that has transpired here, and then of course, the more local and territorial, and then global history. We really should prioritize the *layers* of truth, how we can ensure we don't repeat atrocious crimes of the past, and how we can learn from wisdom and honor the wisdom of the past. The tagline for my mother's nonprofit, Costanoan Indian Research, is 'Honor the past to shape the future.' I thoroughly believe that."

Not stopping for a moment, Kanyon continues to point out that history is not static, but very much alive and moving, with powerful effects.

"Yes, I blame settler colonialism, outsiders coming in, but these layers have compounded and compacted over the years to create this mess we're in now. People think they can start making decisions about how we got here, but different circles focus on their own topic, talking only about sea level rise and climate change and water quality, and they only want speakers about these topics, and they have to meet an academic standard. But the decision makers or community members need to be talking to the people who are most drastically impacted in that bioregion, and the Indigenous of that land. They're excluding community members who have lived in that region for generations, maybe because those people are not in those formal circles."

While she points to long-standing systemic issues, she also includes limiting emotional factors within current attempts toward positive change, a stifling climate.

"There is hurt and there is pain that should be witnessed and validated. But right now, our White culture validates fragility in decision making. Fragility is given an equal platform, and it gets a chance to dictate. There is also toxic masculine fragility, and the hyper-feminine fragility as well, and this fragility invalidates a perspective that deserves an equal platform, to share how we got here.

"Because fragility has been given a platform in these decisions, I'm seeing hurt Natives who have navigated injustices for generations come to conclusions about White people, saying 'White people suck.' I witness White people who have not been in circles where, because of their skin, they've navigated situations that many people of color and Indigenous community members have. They're not being exposed to ways of navigating uncomfortable situations, so I'm seeing fragile people express their frustration with hurt Natives who are lashing out. Both are extremes, on each end."

Dealing again with the complexities while actively attempting to form solutions to problematic communication through language, she advocates seeing beyond stereotypes, and she admits to exasperating limitations.

"I'm using civility politics to clarify, and it helps us, but it's also problematic because civility politics silences people of color and Indigenous voices. Take a silver-tongued settler defense lawyer, and then have an emboldened, passionate, angry, African American who is distraught about social inequities. Put them in a room and let outsiders witness them, and they will thoroughly dismiss the emotionally expressive female. We validate how people show up in an argument, and we are brought up in an environment that does not allow that expressive female."

Kanyon pauses, smiles, and returns to her task of illuminating the intersecting layers she sees, her "methodology" of circling back and adding dynamics as she goes. She points out that a prime emotional difficulty now lies in the nature of cultural guilt and accountability.

"Again, we need a focus on honoring truth in history, and we have to have civil conversations that both validate the hurt and also help strengthen community members' ability to navigate uncomfortable conversations. I'm in a lot of conversations where the 'White' Western settler perspective is very prevalent, and they say, 'I didn't do anything. Why am I being blamed for the actions of some of my ancestors?' I'm not asking you to bear the weight of all of your ancestors' indiscretions and negative occurrences. What I am asking is that you advocate for those truths to be brought to the surface, and for you to share that we recognize that these occurrences have happened.

"You're not your ancestors, and because your ancestors whipped someone doesn't mean you should be whipped. At the same time, let's think about whether your ancestors had slaves, and that your family tells the story of how they pulled themselves up by the bootstraps. Maybe your family has a lineage of having patents for inventions in farming. We need to think about intellectual property rights if there were slaves or indentured servants who did not have rights or money to invest in a patent, so other people got credit for their creations. A good portion of our methodology, and the different tools that we use, are actually extracted intellectual property rights of Black and Brown community members, yet we never talk about it.

"We have monolithic agricultural endeavors that think they know about their industry, with pride and ego, yet they're damaging the ecosystem, the water table, and adding more carbon to the air, impacting the temperature of the Earth. When you think about methane gas, and how humans have brought in a diet of more animals and more fats, extractive methods and industrial profit-driven focuses, and that we are suffering more health problems, it's all related, but none of these companies are talking to each other. Monsanto managers have secret organic personal gardens, so do they even consume their own chemical products?"

Looping back to her beginning points, she returns to culture. The discussion is a circle, continually returning to how the children are raised, and how we perceive.

"If we were brought up in a culture that honored truth, and that had these pedagogies and methodologies as a bare-minimum mind-set in the young generation, the next generation would say, 'I want to eat locally. I want products that care about the Earth.' Right now we are sucking the Earth dry. An adult can only donate two liters of blood, yet we are extracting oil and water from the Earth in astronomical amounts, absurd amounts, and not being accountable. We're not even giving the Earth a cookie for the supposed 'donation.' We're just taking and taking and taking.

"We're fracking along tectonic plates. Here in California, we are along the San Andreas Fault line, and I've heard elders share dreams and stories about how this mountain range, the fault line, is like a dragon. Down in the Central Valley it's the tail, and up by San Francisco it's the head and the tongue. If the head or the neck moves and the tail moves, even in restless sleep, those communities are going to be devastated. The San Andreas Fault line is 'locked and loaded.'"

Earth and humans, the natural world and social realities are in a constant flow in Kanyon's analysis, with problems and solutions spoken in the same breath, all encompassing "climate change."

"With homelessness, if we didn't focus on profit-driven decisions around housing, we'd decide differently. Nomadic communities are valid, and there can be ways to have healthy nomadic communities, to navigate both waste management and communal value systems, but many focus only on how to house people within the limited perspective, instead of opening it up and saying, 'There are different ways that people can live in our communities and our environments that are valid.' Tiny homes or spaces that work with composting methods give back, but we have reactionary legislative practices, and a primitive governing system."

It's ironic to hear the word "primitive" in this context, and Kanyon quickly expands on how the word fits into the discussion, because words make a difference, and the residual power of "primitive" is ongoing.

"We have a governing system that tries to dehumanize Native peoples, calling us 'primitive,' yet Indigenous peoples have been considering how their actions and their words impact the next generations. So let's focus on humility, on accountable measures, educating our young not only about the histories of the land, but about the communities that have brought resources and methodologies of how we came to be where we are. We need to look respectfully because we need to ensure that we can honor the wisdom of the past, of our elders that helped us get here with amazing practices and procedures that are effective. Also, how do we not repeat atrocious crimes against humanity? And, what devastating occurrences are going to happen to the next generation?"

In light of those questions, Kanyon acknowledges that she operates within a structure not her own, but her sense of obligation includes the people within it, as shown by her immersion in it through her work, and she explains this piece of the puzzle.

"I am not a proud American. I am a proud person who has ancestry that is older than the United States government. I don't like our legislative practices, and I recognize that I benefit from the system and navigate the world like I do. I am a by-product of that, but I'm not proud of the United States' methodology of problem solving or its accountability.

"I've always wondered why my friends from Germany know more about United States history than many here do, and are also quite accountable to their history. It took losing a war for them to take those steps to be comfortable and humble, so is it going to take the United States experiencing a devastating occurrence for that humility to set in? I hope not. We don't have that kind of time! I want people to be inspired, *now*. I am fighting the fight around honoring truth in history because the more information we can learn, the more equipped we can be going forward."

She looks back to her earlier points about change, diversity of thought, listening, and education. After saying she recognizes her own patterns of ADHD and dyslexia, like many young people today, she describes her school experience.

"I didn't see my Native community, my value systems, or my structure of understanding, and I walk this world as a neuro-divergent individual, so I struggled in our education system. Going forward, we need our systems to celebrate diversity in ways where we can interact by respecting differ-ent perspectives, so that we *can* have difficult conversations. We *can* sit at the same table and find information, but our current systems need to shift and shape to our level."

What Kanyon has said to this point weaves through intersections of need, directions taken in the past, and how seeing those directions affects where we might go. Against this backdrop, she pauses, smiles again, and focuses on a prime cultural tool for her. The words roll out of her.

"My mother and my grandmother always taught me that when songs, ceremony, and dancing stop, so does the Earth. I truly believe that. Many Indigenous cultures of the Americas and around the world share origin sto-ries, and how to navigate the world, so we acknowledge our ancestral and elders' truths. If an origin story says, 'We've always been from here on this land,' we can ground, and we can care about the land we are on because we recognize it as the life bringer, the entity that is a part of us, and that our decisions will impact our community.

"I feel lucky that I've witnessed and been part of many ceremonies where songs are the core element, when there are times when we're all aligned, and it feels really beautiful because all of us are there for a reason. We are all praying, and we're not worried about it 'coming back,' because it's a part of everything. We will be rehabilitated, and reciprocal energy will transpire. It's praying for the world, praying for the aunties or praying for a grandma, praying for a young one, praying for the waters, praying for the ancestors, praying for the next generation.

"There are ceremonies where you're giving time and energy and prayers for four days, four nights, with no food, no water. There are ceremonies where you're dancing for hours, where self-care is important, and in Native communities, self-care is important, but if you are able, put your energy toward that ceremony and push through. Get your butt in there and keep dancing and singing. Sometimes we have no voice or our voices are cracking. We're dancing in the sun, our faces are windburned. Our eyes burn, but we push through. When one's praying for community and offering their energy and their prayers, it's beautiful, and it's intense.

"Ceremony, ritual, song, dance are part of human existence for all Indigenous societies, Earth-based spiritual practices, everywhere around the world. All our ancestors had Earth-based spiritual practices and all were indigenous to a land. Over time, many of us traveled, or were brought forcefully, or fled intentionally, and there are layers there, and many of us have mixed lineages.

"And, we *all* can honor our ancestors to the best of our ability. We can all attempt to make humble, accountable decisions to how we impact the world, and honoring truth in history by participating in song, ritual, ceremony, is very important. It is very important that we honor and acknowledge where songs and ceremonies come from. If an auntie or an elder sings a song, we should be accountable to where we learned it. Great-great-grandparents sing the traditional songs and inform the next generation, and then each generation shares to the next generation. If for some reason the song shifts because either someone's vocal presentation has shifted, or the cadence is different, then we can find a way to cite how that changed because of what generation sang it a certain way. And then who taught who? That's how we can be accountable. It's almost like Indigenous copyright, oral narratives having been dismissed by written historical records claiming sole authority over truth."

Through these narratives, "we are also respecting the context in which they need to be respected. This traditional song is only sung at this time, if it's a sunrise song, or if it's a coming-of-age song, if it's a men's song. At the same time, we're thinking about and praying on what those traditional songs were when our ancestors first gifted life to that song and gave it breath. What were those traditional songs when they were first given that opportunity to be present in the world? And, how are we honoring them? We need to set our intentions, and hold that respect, that accountability.

"There are times that ceremony should only be personal. Like in the morning, someone has their practice and it's perfect for them. Then there are layers where we have traditional culture bearers engage in ceremony that they have to be accountable to, meaning they honor their mentor. I'm a California Native; however, I have many Lakota relatives who are pipe carriers, and they honor and practice navigating the world as a pipe carrier. And there are many protocols and procedures that you must honor and respect and navigate. So it's not as simple as 'I'm going to just do this.'

"We need to have these things in our lives. And it's important that we learn how we can navigate in the world with them. How do we gather in the community and sing the same songs? As a California Native around other Indigenous relatives, I've learned how to support songs I've never heard before, how to listen for certain elements. I don't know what the elements mean, but I know how to respect the protocol and procedure. It's a part of listening and navigating with cultural sensitivity, with cultural humility and cultural competency. It's being present.

"It's a way we are in community together, so I believe ceremony is important to all of us, all nations around the world. And I'm really sad that people in the United States don't have a baseline to learn from, a respect with which to navigate."

As only one small example, Kanyon describes people coming to ceremonies where food is served without the means to eat, and these people have been trained to have an entitled mentality, as if they bought a ticket and should be catered to. As someone who organizes food for hundreds at Canyon events, she moves quickly into microsolutions, ways to encourage practical but shaken reciprocal values.

"People are coming to ceremonies, not giving back. They're using it, like a battery recharge, and this isn't traditional, not of our people, and I'm seeing it become very pervasive even in our own community, to the point where I really wish I could find a bowl, a plate that's made out of wood, and upcycled items that make little fork sets, and a little permanent straw, that we can wash, and all of these would be conscious items. We could have it in a little bag and say, 'All right, in this ceremony we're going to provide to you. And there will be other ceremonies. However, here is your gift. Carry this and be accountable to how you show up. Even if you can't give of yourself, offset how you're walking.' I would hope that this would help shift our communal behaviors and take steps toward healing.

She talks about practicality, a training of this protocol, ceremony, and thought. I feel good about this echo of many elders' teachings given at Indian Canyon over the years, and I tell her so.

"It's really important that we think about that, because Indian Canyon serves as a safe haven for Indigenous people in need. I am lucky, having been raised on the land of my ancestors, rooted, sharing songs and culture on the lands where my grandmother's grandmother's grandmothers, and grandfathers, have always been. And it is a space where community can learn and connect, because not many spaces are available.

"We don't have easy means of getting together in ceremony. We can't just go down the street to the park and start a fire and dance all night. We can't put up a lodge anywhere we want, especially with fires, but fire is spirit. People don't know that place-based ceremonies are really important because they help ground us, connecting to culture and ritual and song and tradition, honoring truth in history in this way, and learning about the layers of why things came to be the way they are."

Beyond our customary interview time now, we chat about interwoven relationships between Native and non-Native. In her interactions with people all over the Bay Area, she tries to be kind, patient, informing, and welcoming. Many groups want Indigenous input, and she attempts to accommodate in all the ways she can. She is quiet now, then, after reflecting

on the question of whether she has anything to add, she jumps to a trou-
bling thing she sees, and her words are on fire. She has to get this out,
against the backdrop of navigating relationships in a good way. It's about
words, and how they form within communities, and it follows her thoughts
about ceremony.

"I'm extremely triggered and frustrated by the word 'shaman,' a word that
comes from the Evenk language of the Sakha territory in Siberia, a word
to identify healers and people who carry medicine and navigate the world
with a certain spiritual and cultural construct. It's not our word. But there
are many books written saying 'Native American shamanism,' and there
are organizations throughout the Bay Area, ironically established in 1977
and 1978, right before Natives had the right to practice religion freely, and
they are totally tokenizing."

She says people started using the word during the civil rights movements,
when Native leaders were reaching out to make genuinely helping connections, and the irony deepens, along with her anger, because commodifica-
tion of Native words is a long-standing way of trivializing Native peoples,
of "owning" them, like "owning" the land.

"The elders sharing the wisdom with the people was not wrong. However, when people came back to the Bay Area and started a business from
what they learned, it was problematic and capitalistic. It's frustrating
because it's saying, 'Oh, I see similarities between this Native group and
this Indigenous group over here,' therefore saying 'shamanism' equals 'this.'
It equals drum rhythm. It equals ceremonies of alignment of this frequency.
It equals totem animals. They're teaching workshops on how to 'become
your own shaman,' and they shouldn't be doing this."

She illuminates that the sale of supposed Native practices causes political and social damage, and there is a long list of people doing this in the
Bay Area in many forms. Native peoples struggling for rights over sacred
sites or positive school curriculum are hampered in the minds of the public when their "religion" or lifeways are "for sale" in a weekly newspaper
ad. This is not benign. It holds back real relationships with real Native
communities.

"That is exactly why they're problematic. I don't have a problem with the people who are seeking. When I see non-Indigenous people gravitating to Indigenous spirituality, I believe it's their own ancestors calling them home. It's their own genetic memory, wanting them to reconnect to the Earth-based spiritual practices. But because we are brought up in a materialistic society that feels entitled to extractive methodologies, people do not know how to navigate a social and cultural environment like that. They are seeking something, and they are thirsty for that knowledge. But I have a *personal* problem with people capitalizing on it.

"It's perpetuating settler violence and a mentality that is exclusionary and of a White supremacist culture. It's problematic emotionally, spiritually, economically, and it is immoral."

Her anger flares, then she moves to a softer voice still firm with resolve. "I've always heard from Dr. Babe [Dr. Darryl Babe Wilson] that 'America is spiritually, morally, and ethically bankrupt,'" and this is still another example. But even with the disrespect shown by commodifying, she tries to explain how her culture works, coming back to her purpose in navigating an uncomfortable conversation.

"It is wrong to charge for ceremony. I believe that. However, reciprocal relationships occur within ceremonial sharing if there is someone who is a medicine carrier. But it should be reciprocated through energy, time, and effort, and as much as one can afford should be offered, so the person who is a medicine carrier can just do their job."

She recounts what she's been taught by her elders, that "traditionally, if you're a community member who has a child who is ill, the person who provides the medicine is not going to say, 'I won't help your child unless you pay me this much.' They would do as much as possible to help to the best of their ability. And when that child starts to get well, what is it worth to a parent who will then offer as much as they can to provide for that doctor, in gratitude?

"So yes, I despise the word shaman. It feels disrespectful. It is not humble. And, it's problematic in the long run because it perpetuates not seeing Indigenous peoples as valued community members that are knowledgeable, to share with, and to be a resource to the *whole* community. It also does not acknowledge the many cultures and languages of diverse Indigenous communities that have their own 'doctors,' 'healers,' 'suckers,' and more."

She quiets as she says these final words. Navigating how to be of help is the heart of her contribution, like her elders and culture are a resource to her, and as they've taught her. Even in her deep anger at still another layer of the taking and taking, she circles back to this intention. She has concluded this discussion with an "uncomfortable conversation," and it now feels complete. No matter what, she will continue singing the grandmother-honoring song for a very long time. It's been handed down to her, as it came to her, and she knew it was meant to be shared with her community.

16

The Honorable Ron W. Goode
(*North Fork Mono*)

Restoration

COMPOSED BY DAHR JAMAIL

The first thing we do is our ceremonial blessing ceremony for our work. We want to connect to the land, want to connect to the spirits of the land. We want them to know why we're here. They don't care who I am, or what my name is. They don't care where I come from. They want to know why I'm here and what I plan to do. Just because you're Indian, that doesn't mean you know what the hell you're doing. So we get questioned by the spirits. Then you sing a good song. Sometimes you just sing the song, and then you don't go anywhere. It doesn't travel. It's dead. It's dead because either everything's absorbing it or they're just letting you sing by yourself. You sing two to three verses and pretty soon here comes Butterfly. Here comes Dragonfly. Spider comes out of his hole to start listening. Birds come, by your third verse. They're chirping with you. And now the trees are hearing your voice. By the time you're done, your voice is traveling throughout the forest or over the land because now they're accepting you. Now they're hearing why you're here. They understand. So now you're going to be helped by them. This is being connected.

—The Honorable Ron W. Goode

Ron Goode lives for fire. His work centers around Native-led cultural burns in the region of California where he lives. Even the uninformed outsider can see, through the lens of runaway wildfire seasons in California each year, that the imperative for this has never been as great. It is called "cultural burning," because, as Ron told us, "we cultivate."

"The land is hungry, hungry for the return of traditions and traditional ways; hungry for 'proper fire' back on the land, hungry for the spirituality," Ron has written. "The land, the spirits of the land have been waiting for decades, for centuries, for this 'Ceremonial Fire.'"

Ron's work emphasizes understanding the deep relations between humans, plants, and animals, along with the four elements. This is based upon the fact that we are all community, and depend upon one another, and must work together. Therefore, bringing fire to the land rejuvenates cultural resources, and returns new life and medicines that are essential to *all* species.

A retired American Indian Studies adjunct instructor now living in Clovis, California, Ron has been tribal chairman of the North Fork Mono Tribe since 1983. His people have lived in the Central Sierra region for more than ten thousand years. Having taught for decades, in 2006 he was California Indian Education Teacher of the Year, and that same year as well as the following year was nominated for *Who's Who Among America's Teachers*. A published author, Ron was also the coordinating lead author for the *Summary Report from Tribal and Indigenous Communities for California's Fourth Climate Change Assessment*, and his writing and presentations have gained worldwide attention, given that he takes his practical work of using fire to properly tend the land with his tribal and ecological team and presents it at universities, colleges, seminars, and webinars. Ron is also on the Native American Advisory Committee for the Department of Water Resources for the California Water Plan Update, and a co-founder and former summit chair of the California State Agencies and California Tribes Water Summit for several years. He is also an army veteran with a sixth-degree black belt in judo.

When we spoke with him from his home, Ron began by discussing the challenges of not being a federally recognized tribe.

He pointed out the disparity between how a rancher and his land are federally recognized, but how his tribe, who has been on the same land for more than ten thousand years, is not. And in the process of attempting to gain federal recognition, he pointed out how the federal system is a "stacked deck" against Natives. "For us, you're not considered Indian if you're not federally recognized by the Bureau [of Indian Affairs]. I call it governmental genocide."

We all sat with that for a moment, before asking Ron to discuss his work with fire in the context of how federal and nonfederal recognition affects coalitions between tribes. He began by stating how his tribe obtained formal recognition from the county of Madera in 1989. Because the county is part of the constitution of the state of California, they are formally recognized as a tribe in the county, but not federally, and are a tribe with a tribal nonprofit classification. This local recognition does help, as he further pointed out.

"We're mapping all of our trails and putting our Indian names on them," Ron explained. "We've created a special watershed map with our allotments in our old villages and modern homesteads, all in this watershed to show how the people of this watershed interacted with the archaeology. Part of the relicensing to PG&E is that they have the archaeology of twenty-eight sites, and I asked them, 'Who was there? And how did they get there?'"

Ron elaborated upon increasing the understanding of the area through his knowledge. "So these are the people that were there, and they were at 6,500 sites, which are now estimated to be at least eight thousand years old. So by our developing this watershed map, it shows this fact over time, and the alignment of the villages." Ron's work, along with his tribal members, shows the alignments of the villages dating back to records from 1891.

"So we have all of this that proves how long we've been on this land. This is the evidence. It's fantastic, and we expect it to be a model for people to start looking at. And it's only one watershed. Our homeland is between 1.2 to 1.4 million acres, and this is just one watershed of that homeland, but such an important piece to show where the trails came in or where they went out, who got here and who we traded with. That's the kind of work and projects that we've been doing all this last year, so it's kept us pretty busy. Then we do meadow restoration, where we are the lead people in the state for meadow restoration during the drought time."

Ron and the tribe have restored six meadows thus far, and he described them as "fully recovered." This year he has nine more that he will begin work on, which he expects will take four to five years to bring all the way back. His tribe's work is in conjunction with the Sierra National Forest, with which they signed an agreement in 2018 and are now starting to implement. "It takes a while, but it will allow us to do our own cultural burning, and we will be doing our own certifications and teaching out there. When we work with the Forest Service, they certify their people and then they send them out to us to be trained. We have to retrain them. So we said, 'Stop that; send them to us, we'll train them, we'll certify them.' So that's a part of our process." He added another crucial part as well. "There are a lot of policies that the state and federal governments have around burning. In the state of California, burning is still not allowed without liability insurance."

While there is legislation pending that could allow burn bosses the opportunity to be covered by the state for liability, at the time of this writing they are the ones held liable if the burn goes out of control. It is the same with tribes as it is with ranchers, which is why so many of them no longer burn.

In 2018, California allowed ranchers to burn, as long as CAL FIRE did the work, thus taking on 100 percent of the liability. That tactic aside, in California today, if you want to be a burn boss like Ron, you assume 25 percent of the liability while the state takes on the rest. "But if you're going to pay 25 percent, you'd still better get full coverage because you need that as backup in case your fire escapes," Ron added. "That's where ranchers and tribes and everybody else says, 'Okay, you know, we're not doing it.'"

Ron and his tribe burn annually. They had their first cultural burn of 2021 in February, just before we spoke with him, and their second burn was rapidly approaching. "These take an enormous amount of preparation. We go out, we put in fire lines, and we section off each little burn. So if I'm burning three acres, they are probably divided minimally three times, at least an acre each, or less, maybe even into half acres. Last time we had sixteen burns over all, probably over thirty acres in different spots. We section them all off, then we can burn. I had fifty-five volunteers and they're working two or three hours to prep the site and maybe burn for an hour, but we don't have fire trucks, and none of our people are red-carded. We don't wear little yellow suits. We're all traditional there. We do have firefighters coming to join us from other tribes, from CAL FIRE, from the Forest Service. Even the governor's office has come to a burn with us."

Ron distinguished between different types of burns, from broadcast burns, where an entire pasture of several acres needs to be burned due to dead brush or invasive species, to pile burning that results from clearing trees and taking down dead limbs. "There are all aspects of cultural burning and how you go about getting it done. If we have water, we have little buckets lined out all around our roads. Everybody's got a tool, everybody's got rakes, everybody's got shovels, everybody's got clippers. CAL FIRE comes with some bigger tools, in case we need their support, but in the end, it's just us, and we're putting fire on the land traditionally."

How they burn in each place is also contingent upon what Ron called "air boards," which are basically California's air monitoring/air pollution monitoring guidelines. These guidelines differ from area to area.

"When you're out of Fresno, the air board not only has vehicles to consider, but also cows. They have to factor in agriculture mixed with you trying to burn up in the mountains. That's where the big conflicts come because people down here complain as soon as smoke's in the air, and the agencies all want to burn big, one hundred acres, five hundred acres versus our only burning ten acres. But you've got to consistently burn. You burn all the time. But unfortunately, it costs a lot of money to burn all the time. Just to put on a weekend cultural burn is eight to ten thousand dollars. And that's just to invite your volunteers, not to pay them. But you've got to feed them. You've got to have porta-potties. And you've got your permits, you've got your liability."

That might sound like a lot, but then Ron explained what state and federal agencies require if they are going to do a burn. They require twenty to twenty-five firefighters, three to five fire trucks, and if there are other fires happening, like in the last several summers, they simply can't burn because they don't have the requisite infrastructure. "That's sad in the sense that they're relying too much on their equipment, and not on their sense of how to burn," Ron added. "And that's because they don't prep."

It's important to note that with all the burns Ron and his tribe have conducted, they've never had an injury, nor had a fire escape them. Using crews of between ten and twenty people working three to five months, in addition to having between forty and a hundred volunteers and conducting burns five to six times every year, year after year, they've not had a single injury. The tribal spiritual leader, Keith Turner, told Ron this is because they bless the land, themselves, and ask Creator to watch over them.

Given how severe the wildfires have become in California, we asked Ron if he had seen an increase in interest from state and federal agencies in his work.

"There's an interest in what we're doing, and there's a stronger emphasis on the state and federal agencies putting more fire on the land. They're being pushed to put more fire on the land. And after this particular season [2020], because 4.5 percent of California burned, they see that we have to get more fire on the land. It's too bad that it's taking this kind of a wake-up call for them to get to this point. The governor is pushing more fire on the land, and they've been doing that for the last three or four years, and there are some tribes trying to move forward with more fire. Even locally here where I'm at, a number of the tribes either have us come and burn on the rancherias, or they're starting to do their own burning. So that's been real good, since there are lots of other groups that are forming their own cultural burnings. So I see a push coming."

While this is a positive trend, given the increasing severity in size, intensity, and frequency of California wildfires, we asked Ron what wildfire prevention and proper care of the land using fire should look like. He said he thinks we need to take a broad-based view of the situation, and work our way out from there.

"First of all, I would go to the meadows, because the meadow is the hub of the forest. We have 8 to 10,000 meadows in the Sierra Nevada, 70 percent of them needing maintenance or repair, according to a study by the Forest Service, 20 percent of the other meadows nonexistent, and only 5 percent are healthy. It doesn't mean that they're fully functional, it just means they're healthy. Again, of 10,000 meadows, 20 percent are nonexistent. What that means is that there's some sort of meadow ecosystem still in place, but maybe the water is not there, the springs aren't there, it's overgrown, it's lost its functionality."

Ron explained that what they do in their cultural burns is to come in and eradicate that loss of functionality. Fire needs to be brought to them. "We need to be able to burn, and we haven't been able to, so what we had to do with the six meadows that we did recently was to bring in chippers, and the Forest Service helped by doing that. They brought in big chippers, but that's a lot of money, and a lot of time for teams to come in and be able to do that. So, we put fire on the ground to take care of this. What happens is that too often the public wants to equate fire on the ground with a wildfire roaring through the forest. And that's not how it's supposed to be."

Ron mentioned how, in working with meadows, you are dealing with resource areas, and he got specific about some of these. "So take oak groves, for example. You want to feed the wildlife. Indians gather acorns, but all sorts of wildlife eat acorns and acorn is necessary for new growth. You got to feed the bears. You gotta feed the turkeys. You gotta feed the squirrels, the blue jays, and the woodpeckers. And if you're feeding the squirrel, you're feeding the fisher. If you're feeding the fisher, you're feeding the bobcat. It goes up line. So your oak groves are thirty to sixty folks [trees] in a group and maybe 10 percent are producing acorn when it should be 80 to 90 percent produced in April. So, you have to go in and put fire under those oaks without killing them. You have to prep it. You have to broadcast it to get rid of the wood weevils and the worms and the duff. Then you come back and you put more fire and smoke it. Then it takes two or three years before that tree starts to return. But once it returns healthy, you're taking oak trees that are habitat trees that basically were on their way out in another year or two, and turned them completely around from where they were producing 10 percent acorns to now producing 25 percent."

He cited an example of a burn he did in Sequoia National Park where they had two hundred blue oaks producing 36 percent acorns. After two or three years of burning, they were up to 50 percent, which Ron described as "really good, but still not where it needs to be." He continued to explain limitations in approach: "And then when you ask these agencies, 'What's your list of your wildlife?' they can show you their endangered species and threatened species, but they can't give you a list of every species that's out there, from the ant to the rat to the birds, the butterflies and flies, the bears and deer. They can't tell you what they have. Then how do you know what you're feeding?"

Ron smiled big and paused for effect, holding his hands up to emphasize the question. "You're going to put on a big feed at home and you tell your wife to invite the community. But you got this little barbecue pit out back. How many in the community are coming? You don't know. You don't know how much meat to buy. You don't know anything. How many of these people did you invite? You said, 'Invite the whole community.' Well, that's the way it is in the forest. It's the way it is in our parks. They don't know who comes to dinner, and if you don't know who comes, how do you know how much food you got? So that's all part of the process of where we are for a healthy forest or healthy woodland, or even a healthy valley floor. What are you feeding? Who needs to be fed and how are they getting fed? And if we're still in fire suppression, then we've got problems."

Ron explained more of the intricacies of why burning is so important. When there is too much brush, or the forest canopy is too thick, snow or enough rain cannot come through because they are blocked, so the whole ground is not getting soaked. Instead, the water runs off of the limited areas of exposure, creating a very short water season because all the water runs downstream instead of filtering into the ground.

"So you have to open this land up. When the Native American lived on the land, 350,000 California Indians on the land pre-1850, we burned

2 percent of California, which would have been pretty close to two million acres. This last year [2020], wildfires burned 4.5 percent. In 2008, 2015, and 2018 between 2 and 4 percent of the state burned from wildfires."

Ron mentioned that he was starting a burning program with the Seminoles in Florida the day after we spoke with him. "In Florida, they are burning two million acres a year, and they have different policies. Their whole attitude is different on where they need to burn what they need to burn. When you have that, this is what you can accomplish. We have to get to that point where we're able to take care of the land we have, and we have to up that level. Our urban sprawl is so bad in California, where are you going to put fire that won't escape on you and might take out a community? But on the other hand, when we don't do anything, we can look at the communities that have been taken out since 2015. Whole cities are done. So what's your choice?"

Ron cited the Paradise Fire, which started fourteen miles out of town, then tore through ranches en route to the city since not one of them had a fire line around their properties. "It went past the ranchers and climbed the mountain, then went down into the forest which was too thick, washed right through that and headed right into town. The people who got out, barely got out. It could have been a lot worse than what it was. Their roadways are full with trees. We've been telling them for years, 'All your highways, all your roadways are not fire breaks because your trees lean over the top and they kiss each other. There's nothing to stop the fire from coming here and going through those areas. This road is not good. You need to be able to go here and push everything back.' We need to go back to the Indian way of thinking, that when we lived on the land, we didn't have horses, we didn't have cars, we didn't have fire trucks. We didn't have chain saws. We had fire as a tool."

Ron described the situation when a large fire might come, like from lightning, as well as why and how to be prepared for it. "So he [fire] comes, and then that smoke gets there two or three miles ahead of time, maybe then the heat, half a mile out. But by that time you're done with. You gotta be able to pick up your family, your village, get them to different ground, get onto those trails. You couldn't have just a little pathway because while you're moving your family or your village, so was Mama Bear and her cubs. So was Elk and her cubs. So was Wolf and her cubs. So was Coyote. Everybody's moving their family at the same time. Hopefully nobody's going to stop and have a little lunch snack. But everybody's got to get out. That's part of our problem today with the fires after 150 years of suppression. The deer have no trails. They don't know where to go. Where do they go? They go to hide. So when they go to hide, they get burned up because they don't know how to escape." He paused, then concluded, "Those trails were always kept open."

Citing evidence from early settler observations, Ron mentioned Joseph Kenyon's diaries from the late 1800s, when he would look out of his cabin in what is now Fresno County, and write about seeing seven fires going before the sun rose, and note how "the Indians were burning again, every day, fires at different places and different times."

"So we haven't got to this philosophy yet with the government agencies, to understand that to burn every day means smaller burns, like twenty acres, fifty acres, and your smoke is not going to come down here in the valley and upset people. No, they want to burn a hundred acres, five hundred acres, and they want to burn all week because they have to have their burn infrastructure in place. We have to get the state, whether it be the air board or CAL FIRE, to come up with a cultural burn permit that allows tribes to burn no matter what, unless it's an extreme burn-danger day."

Currently, according to Ron, a burn permit under 300 acres is valid only from January to May. "And so what happens is I have to get whatever burn I can, because everything's changing. Then I look at the ground, so I burned this February. Normally I don't have a lot of rain in February, but usually in March. So I'm going to burn this time again in March, but the grass is already coming up, so I might get a pasture burn, because I can't burn the grass right now. Not because it's green, but because if I kill it now, it won't be up until next year or the year after, and I'm defeating my purpose. So there's a shorter window of when you can burn certain elements. If I waited for April, it's too late to burn that kind of stuff at a lower elevation. I got to move up to my meadows above three thousand feet. Now I can burn there until summer. These are things that we understand as Native people, where and when to put fire on the ground. Normally we used to burn in the fall time, and you can't do that now."

The Forest Service tried to burn during fall 2020 with three firefighter groups, and all three of the burns escaped, even with all of their heavy equipment. While they managed to eventually extinguish the runaway fires, it was a hard lesson. "That's why for us, we can't burn, because it's too big a risk. But that's where we used to burn, because we could count on it, because we did small areas. We could burn gently, in a way that's not a problem. But right now, because there is 150 years of suppression, you can't burn gently, but, you better burn light. You can't burn gently because there is no *gentle* burning anymore."

Ron further acknowledged how the government policy of fire suppression that has been ongoing for so long couples with urban sprawl, which has led to homes being built in marginal places. This factor also makes it an entirely different scenario today.

"So if you go to put a fire seven miles down, and that thing gets away from you, all those homes on that ridge above you are gone, and that's your responsibility. That's why the Forest Service doesn't burn in a good way where they need to be burning. Everybody can blame the Forest Service for not burning, but then who's taking responsibility for building all these homes everywhere? Even in the Creek Fire, for these folks that got burned out, it's very sad, but I visited some of these people. They were very happy about the big sugar pine growing next to their kitchen window. But with the nice, beautiful cedar tree growing up through their deck (because that's why they moved to the mountains, to have that effect), guess what happens? You get a wildfire and you burn. No need to do any crying because that's what's going to happen, because you've planned on it. Your trees are growing around your house. This is how we are living now, so while we'd like to go back and take care of the land like we used to, now it's almost impossible, at least on the larger scale."

So Ron and his burners are left to do their work on a small scale, involving loads of prep work, and that is what they continue to do, even though they do not have full support from the state or federal governments. "While I've got CAL FIRE and Forest Service out there with me, I'm only talking to one or two people, not the agency. It's just one or two people coming and volunteering their time, but it's good because they see what we're doing and they have a better understanding and then go back to their agency and talk about it. But again, they're only one man. I don't care if you're the captain, you're still only one man."

Ron had mentioned working with the four natural elements and holding a "broad view," so we asked him to further detail how fire and water work together, how fire is used to clear the meadows in order to create more water for the whole system.

"When you eradicate what doesn't belong in the meadow, you open it up. You're not going to allow the water suckers to be draining the meadow, so your water table comes up. When you look at the conifers and oaks, they can reach down six to eight feet. Shrubbery can only go three feet. Plants, flowers, they're only one to two feet. They gotta be able to hold that water at the very surface in order for them to survive. That's why they have a short season when they do survive, because they don't have much root system. Whatever has a bigger root system can hold water longer, which means that it's holding the water higher. So when we are burning and we have a rejuvenated plant, it's now fresh, it can hold water in its leaves and it can hold water in its root system. That pulls the water, and they're pulling the water to a higher level. So every time we've taken a meadow that was nonexistent and restored it, cleared it, the springs come back and the water comes back full. When it does that, then pretty soon we've got a meadow. Then we have more species, and different flowers are coming back as well, as Bear walks through from one place to another, as Deer comes through from one place to another, and as Hawk flies in and Crow comes in, and everybody's bringing seed from someplace else."

Ron spoke in a very animated way about all of this; his eyes lit up, and he gestured more with his hands, as though forming the meadow in his mind. "The different animals start coming in. Everybody starts coming back because now you've got a nice meadow. You've got water and sunshine. Once you open it up, you get those ingredients, so now you've got something going on.

"I recently wrote on an article ["A Community of Relations, a Relationship of Connectedness," 2019] on a community of relations between the plants, animals, and people. I'd known this already, but talking to some ranchers when I was doing my trail work this summer, they were telling me how the oaks that are deciduous, when they stop their growth in the fall, they quit drinking water. And at that particular time, you'll start to see the ponds rise, the springs rise, dampness in the creeks, and then maybe even water. Eventually all the water starts rising and coming back on the land. It's there, but it's not being used by everything. The natural system that's in place has its own system of how it's going to operate when it needs water and when it doesn't, and when it allows the water to come back up."

In the article, Ron cites how various species interact and influence one another within a community in ecology. His tribe restored several meadows back to a healthy state from a "nonexistent" phase, during which its functionality no longer existed. After their restoration, "the water tables were brought back to full levels and springs continue to run all year. Species counts on the meadows went from 20 to 120, and as high as 160 different species came back into the meadow once it was cleared and the water table rose. And, this was all accomplished during the last six-year drought in California."

Ron went on to tell us how last year (2020), due to the climate crisis, they didn't get frost at the usual time due to warmer temperatures. That meant the trees kept using water longer, and it wasn't until early November before the acorns began to drop, after which their cycle is complete and the trees go dormant for a couple of months while awaiting the return of the rains. However, if the trees had started dropping their acorns in September and October, then the water would rise in October. "But in this case of the climate changing, the oaks didn't drop their acorns the way they are supposed to. So it all goes hand in hand with what the ecological system has in store. That's what climate change is and how it's affecting the land."

I asked Ron what the average person might do to help Indigenous people upon whose land they are living, given the severity of our current situation. "They have to get active themselves. They can't be complacent, and they can't just sit back and gripe and holler and say the government doesn't know what they're doing. The government can only do what they can with what they've got, and they need more help. And that's where the people come in. Maybe they can't help the government, and maybe they can't help us. But maybe they can help a politician, or maybe they can at least have a better understanding of the philosophy we live by."

Ron has been discussing the issues of philosophy and action for a long time with many people, as a teacher, practitioner of traditional land management, writer, and community leader. He relates that he has watched Western scientists carrying the attitude that they have *the* ability to find a solution to the climate crisis. "Good luck," has been his response. "After listening to some of them again recently, I found out that they weren't connected to the land. They're disconnected. You know, everything was 'Nature.' Everything was, 'Oh, I liked being on the land.' These were the professors talking. They've never been a practitioner, they've never put fire on the land themselves. They don't know where the water comes from. So until you are out there and work with it, as Native Americans do, you don't know. The tribe has the Indigenous knowledge, has the generational knowledge that's been passed down, but having the knowledge and understanding Traditional Ecological Knowledge, TEK, doesn't mean you know what you're doing."

Ron paused, thought for a moment, smiled, then continued. "It's kinda like getting a complicated thing from the store that you got to put together, and you study the instructions and you're like, 'Oh, just throw these damned instructions away. Let me try to put this thing together.' So you come up with extra parts when you get done, but as long as it 'works,' you use it.

That's kind of the way that we are today. But that's not the situation when you're *from* the land. When you're from the land, you understand the land. You have to already know what you're working with. Like I told some of them, 'When you're out on the land, you just know.'"

Ron then went on to describe the process he and those he's working with go through. "The first thing we do is our ceremonial blessing, the ceremony for our work. We want to connect to the land, and to the spirits of the land. We want them to know why we're here. They don't care who I am, or what my name is. They don't care where I come from. They want to know why I'm here and what I plan to do. Just because you're Indian, that doesn't mean you know what the hell you're doing. So we get questioned by the spirits. Then you sing a good song. Sometimes you just sing the song, and you don't go anywhere. It doesn't travel. It's dead. It's dead because either everything's absorbing it or they're just letting you sing by yourself. You sing two to three verses and pretty soon here comes Butterfly. Here comes Dragonfly. Spider comes out of his hole to start listening. Birds come, by your third verse. They're chirping with you. And now the trees are hearing your voice. By the time you're done, your voice is traveling throughout the forest and over the land because now they're accepting you. Now they're hearing why you're here. They understand. So now you're going to be helped by them. This is being connected."

Ron paused, and we all sat together in silence, smiling, feeling all of his words.

"As I told these professors yesterday, you use all this fancy college jargon and you tell me how you're connected. You have a computer and you have a plug, and this is what you're doing. You're running around looking for a plug-in and you're running around trying to find Wi-Fi so you can be connected. If you're Indigenous and you're on the land, you're already connected. We already understand where the Wi-Fi is.

"But you have to try to speak the lingo to these people so that they can understand, because many of them talk about Indigenous people. Yesterday they were acknowledging the Indigenous leaders, but said not one word about practitioners. Some of these Indigenous leaders, they don't know anything about the environment. To me, we keep hearing the cliché of TEK. We hear people say, 'Oh, TEK, that's spiritual,' but I've not seen them use it."

At that Ron smiled, challenging the surrounding systems, but in order to teach, to say what is needed. Continuing to speak as though he was speaking to the scientists, he went on.

"You believe in my philosophy, my spirituality? You understand it? You understand what I'm saying? Do you understand that? That the land has a spirit and everything has the same spirit, and that everything out there is my relative? If you believe that, then I guess you can use TEK. But you don't, because I highly doubt that your religion thinks that way. It thinks that everything's separate. There's us, and then there's animals. There's us, and then there's the land. There's us, and then there's everything else. Most of your religions say that people were put here to have dominion over everything. That's why the government agencies think that they're managers, land managers, and managers of the wildlife. But the first thing you gotta do is throw out the word manager. You don't know how to manage. Secondly, you need to get rid of dominion because there is no dominion."

He waited a moment then spoke again. "If you don't understand your land, you won't understand the plant. And if you don't understand the animal, then you're not connected."

Ron reflected on hearing "religious" people talk about having had dream interpreters a long time ago, "during Joseph's days," he said. "But we don't have dream interpreters anymore, really. Back in Adam and Eve's days, they used to talk to the animals, but they don't do that anymore. Somewhere along the line, somebody lost something."

He smiled again. "You still *can*, and *that's* TEK."

17

Corrina Gould
(*Confederated Villages of Lisjan*)
Recognition

COMPOSED BY STAN RUSHWORTH

> I think that if we focus on the kindness of people, we can get a lot done.
>
> —Corrina Gould

Corrina Gould is an incredibly busy woman, so we were deeply grateful that she carved out enough time to speak with us. Despite COVID, she is thoroughly immersed in organizing, teaching, and participating in events advocating for Native peoples in the San Francisco Bay Area and beyond.

An online description of her work with the Sogorea Te' Land Trust details just how much she is involved with, as a Confederated Villages of Lisjan woman born and raised in Oakland, California, or the village of Huichin. She is the co-founder and a lead organizer for Indian People Organizing for Change (IPOC), a small Native-run group that works on Indigenous peoples' issues. Further, she is the spokesperson for the Confederated Villages of Lisjan/Ohlone, and co-director of the Sogorea Te' Land Trust, the first Indigenous women–led land trust in the nation.

After I watched *Beyond Recognition*, a film about Corrina and her colleagues' work, two things stood out that I wanted to learn more about, both vital issues to Native communities in different ways. The first is the question of federal recognition of tribal groups, which can be a deeply frustrating question for many. The second, which is very much connected, is the education of both Native and non-Native children. Education helps children know who they are and who their neighbors are, through history, language, and intercultural exchange, so both questions circle around personal and community identity, and the far-reaching implications Corrina is concerned with.

Corrina smiles and nods, then begins. As she does so, an image from the film stays in my mind. She is standing in front of a large crowd in downtown Oakland with her clapper stick, talking about what needs to be done toward responsibility. She is wearing a silver and black Oakland Raiders jacket, and it's blowing in a light breeze. She is strong, sure, and kind. She encompasses all the people in the crowd, including them all with her gaze, in her city, on her ancestral ground. She brings everyone home, in the best way possible. Part of gathering fellow citizens together involves knowing what issues Native people are dealing with, here and now, and in that spirit, she dives into my question.

"All right. Where do we start? Let's start with *Beyond Recognition*, with questions about federal recognition versus nonfederal recognition.

"One of the things happening when I was younger, was that there was a group of Ohlone people trying to get federally recognized, and we were part of that group for a while. That particular tribal group got denied twice, and we realized there was no way the federal recognition was going to go through."

Corrina explains the difficulty of working within what she terms "racist" recognition policies for people in California who have been "dealing with three waves of genocide consecutively: the Spanish, Mexican, and American periods." The policies require that California Indian communities prove they held their cultures together throughout those periods, despite the attempts to destroy them.

She pauses, then reflects upon what is most important to her, thoughts that are obviously the result of years of careful consideration and experience. She states her essential purpose in a key question: "How do you begin to do the work of healing?"

Describing the road to federal recognition, she lays out a process of "trying to raise money for lawyers and researchers and anthropologists and ethnographers to put together everything, the way of federal recognition," then asks, "Is that the real work?

"I had to say that that's not the real work for me. I am recognized by my ancestors. I was born to the land that I have always been on. My ancestors have always been on that land, and we have had a connection to it since the beginning of time. And as long as I recognize my ancestors and they recognize me, the work can happen."

She expands her thoughts with a strong smile of gratitude. "I truly believe that's why the work that we've created here in the Bay Area has been so abundant. I believe this. I talk to people about the abundance that was always in the Bay Area before colonization," she says, and this is a new abundance.

Regarding problems involved with federal recognition for Bay Area peoples, she points to important contrasts between California and many other specific parts of the country. In her homeland, "colonization only happened a couple of hundred years ago. It's not like the East Coast. Our colonization was around the same time the Declaration of Independence was happening, when the Spanish missions started coming up to this area, so it looks different, and the practices of the federal government in terms of recognition favor the tribes coming forward on the East Coast, not tribes that dealt with the Spanish missions and Mexican Rancho period, and then the annihilation by the U.S. government.

"It's difficult to prove the things they want us to prove, having a government set up with tribal roll numbers and minutes and all those things they require, when we were running from genocide." She also brings it to the difficulties of meeting government requirements on an everyday basis. "It's almost an impossibility to do while raising children in the Bay Area, here in our homelands." Again, she points to how the difficulties of the process brought her to another resolve.

"This made me look and see what it is that we really needed, at a time when we really needed to do the work of recognizing and honoring our ancestors and telling our story." She brings a sense of urgency to the forefront with a reminder that "it's really only been in my lifetime that the story has been able to be told.

"Our elders, for the most part, kept it really quiet, or it was not well publicized, about Ohlone still being here. I grew up with people saying 'You're all dead.' In fourth-grade history, here where my children grew up, they were still having to put together a mission with little sugar cubes. And I fought those teachers about that."

I picture all she describes, including the teachers facing her, and I know she prevailed, as her priorities are strong and clear. To drive home her larger point, she presents the mathematics of her situation.

"The recognition process is about how to spend time. With hundreds of tribes across the country trying to be recognized by the federal government, and at a process of only 1.5 being recognized per year, if we were to put our hat down at the end of that list, it would be beyond my lifetime before we even got through the process.

"Would that be a good way to spend my life, or would it be better to actually do the work that we're doing? To honor the ancestors, to bring back the cultures, the songs, the dances for the healing that needs to happen here, to begin to trace back that historical trauma and really begin to heal from it, to bring back a people that had almost disappeared? Wouldn't that be better time spent in my life?

"As long as I know who I am, and my ancestors recognize who I am, the work gets done. I don't need a government that took over our land to recognize who I am."

There is much more to think about when considering Corrina's approach to this situation, and after reflecting a moment, she points out how "it's a farce that American Indians are the only ones that have to prove who they say they are in this country. People come from all over the world and never have to prove who they say they are. And yet we do. And we are the original people here. This idea is part of the education system, and it's used to wipe us out consecutively and continuously. I call it a paper genocide that continues to happen.

"As Ohlone, we're taught about in the fourth grade and then we're never spoken about again, and that continues the erasure of our people on our own lands." Corrina reasserts her purpose in light of this deep invisibility. "I tell this story so that my grandchildren don't have to tell these stories, and so there is a way for us to go forward and start healing."

Within that process, some changes in government are happening, which she points to with both consternation and gratitude. "Our tribe had to get on a list for the Native American Heritage Commission, something that I had not chosen to do for many years because I had in my mind that the people that were on that list assisted developers in destroying our sacred places in some kind of way, but there were laws changed that *forced* tribes to get on that list in order for us to legally have a right to say anything. We got on the list to have conversations, to be in consultation about development that's happening in our territory."

As a result, and despite not being federally recognized, her group has "tribal members answering letters and emails and going to site visits and being in meetings about all these buildings that are going up in our territories, ensuring that they won't destroy any sacred sites."

She continues, shaking her head, describing a current action of having to "clean up a mess of a developer destroying a cemetery right now." She pauses, and says with a smile, "Those were the kinds of things I didn't want to be involved in. And yet we find ourselves there. But still, we now have a way of saying, 'This is a sacred place. There are sacred places, and this is how we're going to deal with them.'"

This kind of recognition has evolved from the development of her own community in ways she and her tribal members have defined, nothing solely bureaucratic from the outside. And it's working.

"I find myself in a better position sometimes because we're on that list." An example of what she and her group accomplish through such collaboration is the effort to stop a housing development on the West Berkeley Ohlone shellmound, a 5,700-year-old sacred site. They work together with the city of Berkeley against State of California rulings that override local efforts to protect one of the earliest known Ohlone settlements. In 2020, the National Trust for Historic Preservation named it an "endangered historic place," one of the top eleven such sites in the nation. Building community in this way bears fruit for everyone, so she is resolute. "You have to always find the good with the bad, right?"

She collects all that has been happening into a critical shift in awareness. "The work that we've done over the last few decades in the Bay Area, where people now recognize the Ohlone people, has changed the landscape in terms of recognizing the First Nations people.

"It's been amazing, and it's changed the psyches and hearts of the tribal people that are here that were invisibilized. And, this work is important not just for Ohlone people, but for everybody living in our territory. It's important for people to know where they stand in the world and what land they're on, what the history of that land is, and what their connection to that land is."

These thoughts bring up key questions, and Corrina asks them without hesitation, inviting all citizens to consider them. She was raised in a giant city on her ancestral homeland, so she knows who she is addressing.

"What is your responsibility as a settler? When you sit on someone else's land, a guest, what is your reciprocity with the land and the people? That's the work we have to do. When we talk about climate change, I have lots of thoughts of how we can move forward as a group." She hesitates at the word "forward," then quickly explains that "it's really moving backward."

She explains. "I feel this when I know that human beings lived in smaller places, shared food and had everything they needed within walking distance, and today, biking distance." She elaborates her vision of the future. "I see our cities getting smaller, and human beings living more communally, in a way that our ancestors lived for thousands of years, so that there is an abundance again.

"I have great hope and faith because of my four grandchildren. I think that we as human beings have the ideas, and we have the creativity to clean our waterways, to ensure that there's food and housing for everybody. All the answers we need here on the Earth are already here."

She describes further solutions clearly: "What we need to do is to strip away the systems that no longer serve us, those systems that allow only a few people to hold land, and only a few people to make money, and only a few people to have the ability to actually have stable housing."

As I listen, I imagine readers thinking that these ideas are abstract goals, but it's imperative to remember that Corrina's traditions and thoughts are inherited from a very recent reality.

"I often think about my ancestors' homelands, our homelands now. Two hundred years ago, there was no concept of hunger or homelessness, just two hundred years ago. There was not any concept of it. You could eat out of and drink the water from every fresh creek. And this is not that long ago. We're not talking about thousands of years ago, but two hundred short years ago!"

She lets this sobering number sink in, as we see how long a more human way of living existed. California Indians lived here successfully for millennia, so she asks us to balance these numbers with what is possible. "I don't think it's going to take us two hundred years to reverse that. It's just going to take some willingness to say that it's all of ours.

"I think of the creation story that a lot of Native people have about how human beings were the last ones to be born, the last ones to be created in this circle of creation, and that human beings have taken themselves outside of creation, outside the circle, and have messed things up by saying that everybody and everything else that was created before us, is something less than us. We have to think about this in order to feel this Earth, and to come around, for humans to survive."

She takes a breath, then her thoughts roll out steadily. "The Earth will survive. She has a great way of healing herself. We saw that during COVID when she put us on 'sit time.' In that time, she was able to clear the airs and places where they had pollution. Our relatives in the waters came and swam in places where they hadn't been in a long time. Our coyotes came back into the city of San Francisco, and when we were sitting down, the Earth quickly began to heal herself. If humans can see that as a lesson from this great, horrible disease that was brought here, maybe we can begin to understand that we are the ones that have influenced the problems in this climate disaster. I don't call it a 'change.' It's a disaster, and it's rolling up on us.

"I look at the fires that have happened in California, and all the Native folks are talking about how fire had been a relative we worked with for many years. Diary entries by folks that came here talk about how our lands were clear. There was nothing underneath, no brush. The lands were pristine, and they had been taken care of in a good way, and that took thousands of years of living with the land, knowing the land."

Again, her reminder of the longevity of California Indian stewardship confronts the notion of the "inevitability" of negative human impact. More, it's a manner of approach, a change she addresses.

"The people who showed up here changed waterways and brought animals that destroyed landscapes, and they forgot that they are part of the land. They forgot the responsibilities of human beings to a landscape."

But Corrina affirms again that many have not forgotten. "These are our responsibilities as California Native people. We started those fires not just to clear out brush, but also to bring up and honor the medicines that grow because of those fires, and this honors the animals that come and eat those shoots, and it honors our basketry materials, and it honors all of those things because we work in a circle of reciprocity, and those circles were broken. And, it's time to fix those circles again."

This brings us to the Sogorea Te' Land Trust, and the quarter-acre property that her group acquired. I ask her how it feels to be able to work that land in the middle of this vast city, and how this came about. She tells the long and important story.

"The land trust was created out of a need, and is the first urban Indigenous women–led land trust in the country." She says this with a smile of satisfaction, not only her own in a personal way, but in a much larger sense.

She recalls the beginning: "Johnella LaRose and I co-founded Indian People Organizing for Change during what we call the dot-com era. During that time many people were gentrified out of the area. People were out-bid for their homes and apartments, so waves of people left. And at that time there was also a new wave of development, and that's when our an-cestors started showing up a lot more, because they were building in places where our sacred sites were, and on our burial sites."

Corrina is referring to a very recent spurt of growth, one that necessi-tated a concentrated response addressing the lack of knowledge about sa-cred sites and California history, as well as the immediately destructive results of development.

"Johnella LaRose and I worked with people from all over the world for many years, joining walks and educating people about shellmounds, because twenty years ago nobody in the Bay Area knew what shell-mounds were. It's not taught about in schools, and the regular folks living here just didn't know what the heck that meant. It was really an educa-tional process." The shellmounds are places of ceremony and burials, cre-ated over countless generations, but no one living around them knew this, as they have been part of the erasure of Bay Area Native peoples.

Along with education, more physical actions had to happen as well, because the violations were ongoing and imminent. "On April 14, 2011, we took over a sacred site with Wounded Knee Del Campo, a Milwaukee man who had been trying to stop a development here along the Carquinez Strait for twelve years, one that has two shellmounds on it."

Corrina looks back in time as she continues, reminding us of the differ-ent waves of occupation, their effects, and different forms of Native response.

"It was one of the last village sites where our ancestors held out, and they actually didn't get pulled into the missions from there until 1810. So, on that ancient village site, Sogorea Te', we held the land for 109 days with an intertribal group of people, including folks from all walks of life who re-connected with the land through ceremony and song, and through living in a village, like we had not done for many, many years, during the long silence. And a lot of magic happened there. There were a lot of miracles that happened there."

She details the coalitions formed, different elements working toward a common goal, and points out how communities work when intentions are clear. "It was the first piece of land in the Bay Area, in the country, that had a cultural easement created between a park district, a city, and two federally recognized tribes. Those two tribes came together and pitched in money in order to have equal access to that land, and that saved it from being destroyed." Within those coalitions, however, the inclinations of the settler rules can still be seen. "But in the process of doing that, they signed a contract saying that no more than ten Indians would ever gather at that site, and that there would be no big drums allowed on that site."

While this agreement is a reminder of the limitations often imposed in dealing with government, Corrina develops the point that another force is more powerful.

"Of course, we were angry when we saw the contract, until we realized that we didn't sign it, so we show up every year with hundreds of people and have ceremony there without a permit. We light a fire and do what we're supposed to do because that's our sovereign right, and sovereignty is not something that's given to you by another government. Sovereignty is doing what is your God-given right on this Earth. That is our sovereign right, so we show up and have our ceremonies there." She nods, "And they don't bother us."

She smiles broadly, respectful and confident of the deeper work being done in such coalitions, affirming that everyone knows what is right when they look at the bigger picture, which takes time and work to accomplish.

"This was after we had been walking the shellmounds for four years, from 2005 to 2008. The archaeologist Nels Nelson created a map of 425 shellmounds in 1909, and we decided to do this education and the prayer walk, so we walked from Sogorea Te' all the way down to San Jose and up to San Francisco. For years, we did that with people from all walks of life, with people from Australia and the Cape Verde Islands and Japan and Nova Scotia, and all over the country.

"We walked eighteen miles a day, stopping at each of the shellmounds and laying down prayers in remembrance of our ancestors, and that's when things started happening differently, when we did those prayers around the bay. People now have it instilled in their minds that our ancestors are there. They can say, 'There's a village site there.'"

Corrina gives deep acknowledgment to the connections made with others strictly by Native visibility. She smiles as she tells how that happened. "We're human beings and human beings move, so we walked three hundred miles in three weeks around the entire bay, just like our ancestors had. It was the idea of seeing that landscape in a different way as you're praying, and that was a way of building community, and being in conversation with people."

Then she pauses in her story, coming to how the years of work and gathering people brought her group to taking another next step in relationship to the surrounding society. "When it culminated in taking over that site in 2011, we realized that we were missing some tools."

She laughs out loud, referring to still another "miracle." "So there was our good friend, who is also on the board of the Sogorea Te' Land Trust, Beth Rose Middleton." Dr. Middleton is an assistant professor at the University of California, Davis, focusing on Native environmental policies and activism for the protection of sacred sites, and the author of *Trust in the Land: New Directions in Tribal Conservation.*

"Beth invited me to a conference on Native land trusts in Southern California, and I had no idea what a land trust was. But I went, and I met a handful of people that were joined in land trusts. Some were federally recognized and some were not. Some were casino tribes, but all were Native people trying to get access to their land. I heard stories of people buying back sacred pieces of land that were a part of their tribes, so they could reengage their tribal members in the sacred places they had been cut off from for many years, because they were under private ownership."

At this point, she laughs, recalling that almost all the people at the conference were men. During a lunch, she asked one of them, "So is this a boy's club?" He said, "Yeah, kind of. Not just Native land trusts, but all land trusts are run by men." She found this "interesting," she said, and she and Johnella began discussing the whole idea.

"Because we're not a federally recognized tribe, we don't have a land base, and we've been praying for the ancestors to come home." She elaborates and explains, revealing staggering numbers. "UC Berkeley itself has over 9,000 of our ancestral remains and funerary objects. If they were to give us those ancestors, where are we going to put them? So we said, 'Okay, maybe this is the tool we've been missing. Maybe this is what we need.' We began the journey of creating this land trust, and over the years of creating and naming it, we decided on Sogorea Te', because that is where it was dreamed up.

"Then we began to talk about the conversation I had at the conference. What did that mean? We started making a correlation, relating to what happened across the world with men in charge of the planet, and what happened to the land, the sacredness being taken from the land, and what happened to women's bodies, and the sacredness of women, and the rape of women, and the raping of the land and the total devastation and destruction.

"We started talking about what the land really was and who women really were. We decided it needed to be an urban Indigenous women–led land trust, to bring the balance back into the world that has been missing since men have held on to the land and have been in charge of it."

Corrina is quick to affirm that this is not a negation of men but a rebalancing, an inclusive direction in the face of imbalance. "It doesn't mean that we don't work with men. I have sons and grandsons, and so does Johnella, and they work with us on the land. And we've hired other young men and young women, and two spirit people and artists, and we're trying to envision what our responsibility is. What is the language of the Earth, and how have we misspoken that language? How has it been taken away from us as Indigenous people? How do we come back to the reciprocity? How do we return the language? How do we return the ceremonies? It's about how to return the songs, how to return to ourselves, to our sacred responsibilities to the Earth and our own bodies.

"When colonization happened, it didn't just happen to the Earth. This happened to me, to women, and it happened to men as well. The sacred responsibilities were taken away from men. There was the idea of heteropatriarchy and misogyny, and all of these things thrown in that everyone had to follow in order to survive, and it has been a great devastation to our tribal people and our families. And now it's time to bring that balance back.

"Women have songs for the water and for our basket materials, ceremonies for putting our children's umbilical cords into the land so there's always a connection. It's women's work to not only bring life in, but to see life leave, and we have those ceremonies as we put people down to rest. The imbalance that's happening needs to be changed back into balance, and there needs to be a place for women to do that, to save us as people. So, we became an urban Indigenous women–led land trust, on our traditional territory."

Corrina points out still another important part of the rebalancing, referring to changes in the population of her homeland that must be included. "Our territory was used as the vault for the forced relocation of many Native people during the fifties and sixties. People were pushed into the Bay Area from reservations as a part of the assimilation process of the United States government, hoping that within a few generations, because of blood quantum, they would have erased Indian populations. And it didn't happen. It backfired, just like every other assimilation practice they've tried."

She describes the depth of connection between women of many tribal backgrounds. "I have been a doula with many different Native women, at their children's birth, and we've all raised our children together. Johnella and I raised our kids together, and so there's an intertribal group of folks in the Bay Area, with many who've never gone home again, and they have to have a place too."

She stops, gathers her hands together, and brings us back full circle.

"So this land that we're talking about . . ." she begins, then stops. She reminds us that everything she's said is needed in order "to talk about this quarter acre of land," and the many people to whom it is important.

She tells the story of the young couple who formed Planting Justice, a nonprofit organic nursery working with formerly incarcerated men and women. They went to Standing Rock in 2016, and before they left, they asked the elders what they could do when they went home. "The elders told them they should work with the First Nations people on whose land they live. And they took it to heart." A mutual friend and elder to the couple's group put them together.

"So we met them on their two-acre lot in Oakland, under a California walnut tree at the back of the property, where they have a beautiful organic garden. They told us about the elders at Standing Rock and asked us if we would accept the quarter-acre lot they hadn't started to use yet, that they wanted to return."

Corrina stops here, picturing the land before she goes on. "This land runs along Lisjan Creek, named after our people. It runs along the 880 freeway and butts up against property that's just about all on fill. That's where the bay would have originally been, so the land is actually either on or close to an original village site of my ancestors. It's also only a half-mile walk from my home," she laughs. "So these ancestors' miracles continue to happen.

"We said 'Yes' to the land. And I cry when I go to the land. Almost every time I go onto it."

Without hesitating, she explains still another key part of stewardship that began immediately. "But we could not accept this land without a relationship with these folks. We'd walk past them working the nursery to get to the land, and what does that mean without a relationship with the people that are there? So we decided to take that first year to clean the land and to really only work on building the relationship. And that's what happened."

Corrina is glowing while telling the story, savoring the details.

"And the land began to transform. We put California native plants on it, we cleared it, and then Johnella said, 'Let's put an arbor up.' We looked all over the Bay Area for the redwoods we would need, seventeen feet high, and a lot of them, but we couldn't find them. And finally we found someone who had land up in Sonoma County. They needed it thinned out, so we went up there, ten of us. None of us were tree cutters. We had U-Hauls to carry them home in, and we had no idea what we were doing, except for a young man, Billy, an African American man who worked at Planting Justice. He had been on the fire line during his time in prison, and he talked to the trees. He really talked to the trees, and we asked him if he would lead our crew. We brought our elder, Wounded Knee, and we prayed at each of those trees, asking them to give their lives so that we could re-create this arbor.

"We did it, and nobody got hurt. We cut them all to size and rolled them up onto the U- Hauls and brought them home. They were seven or eight hundred or a thousand pounds each, full of water, and we said, 'Oh, we'll put it up in a month.' And those logs kicked our butts for a year, which was good, because it allowed us to build a different kind of community.

"We invited people to come and help debark and sand them, and to ready the land for them. It took a year, and two years ago this May we put up that arbor. And we put it up intentionally, with a pole for the men and a pole for the women, a pole for the elders and a pole for the little ones, the young ones, and a pole for the two-spirit people. Then all the other poles went up, with the fire in the center."

She has a look of deep satisfaction, then sobers as she continues. "We were going to dance it in last year. We had thousands of white sage and tobacco plants we started from seeds that we grew as part of our giveaway, and then COVID hit. And we weren't able to dance it in, not in that way."

She looks down, then back up at us, pointing out that the arbor continues to do its work despite the pandemic. "We gave away those plants, and even during COVID, we were able to light that fire for one of our good friends, a board member who passed away. We had a fire for her for four days. My daughter got married in that arbor, and this past summer a young man came out of prison and asked permission to come and pray at that arbor before he went home to his reservation. So the arbor is calling people in different ways, and it's a beautiful blessing.

"We've created something called Himmetka there, a place where we all gather. We have fresh water tanks there. We have a rain catchment system there. We're building an outdoor kitchen deck. It's a place where we have food and medicine and first-aid kits. Because this is in a very poor neighborhood, it's one of those places where nobody's going to come save us if there's a natural or human-made disaster. Our responsibility on our land is to take care of our people and our neighbors, so we want to make sure that we have something there. That's part of the work that we've done during COVID, to get food and take it to elders and people with young kids and those that are immune-suppressed, and those that are working on front lines, to make sure that they have good food. It's our responsibility as Indigenous people to do that, to make sure that we feed the people."

She continues to expand on the responsibilities that connect to the quarter acre of land. "It's also to make sure that we have places that are safe for our kids and our neighbors' kids, and to work in reciprocity." She tells how the people from Planting Justice protect the land, and how they all work together, seeing the long term. "That place means everything to me, for my children to go there, and my grandchildren, to be able to go to a place for us to have ceremony.

"When we talk about remaking relations, it's a place for us to come back together in alignment, do the songs and the dances and the ceremonies without outside interference. We didn't have that before we had this piece of the plan. We had to ask permission on our own territory, to go onto lands that we have been connected to for thousands of years.

"But this piece, when we go onto it, it's ours. And no one can tell us to leave. No one can tell us when we can have ceremony and when we can't have ceremony. It's a place that gives to us, and it holds us up to do the work that we're supposed to join in."

As I listen to Corrina, I see the depth of feeling when she talks about the children and the land, that all of this is about the generations for her, so I ask her to talk about education, the second question stemming from her work, and about what she sees the system doing, how it affects young minds, and what needs to happen.

"Because I have kids, I understand what they have to go through." To counter what she knew is happening to them, she went to work. "I worked with a museum for about eight years, and I've done presentations to thousands of little kids." She says she tells history as a simple matter of fact.

"Kids always understood slavery was not a good thing, right? They have a real sense of justice at that age, about what is right and wrong. There are no gray areas in there." In her lessons, she tells the kids of the laws and other difficulties put onto Native people with the successive colonial waves, and while she does this, she observes that "the adults sitting in the back, and the teacher, cringe while the kids would just roll with it. The kids have answers like 'No, it's not good.' I just say, 'You know, this is what happened and it was wrong and it was scary. And then this is what's happening now,' and in a matter-of-fact way, I tell them the information, because it doesn't harm them.

"Leaving it out harms them because when you don't tell them the information, when they're seventeen and eighteen years old, they're having to deal with the truth, and the adults in their lives who withheld the truth. And why?

"They're not fragile, you know. We have to see that the kids are more fragile without having the truth to hold them up. It makes it difficult to have conversations with adults who have been told the lies for all these years, and we end up with a country that is split the way it is, that doesn't want to deal with truth. We have to be able to set ourselves free from those lies in order for us to move forward.

"Kids absolutely understand that, and it's horrific for us to hold on to the lies. They have to find out when they're younger, or they will not find out at all. It's our responsibility as primary educators to give the truth to the young people and bring their parents along with them. Young kids go home and have conversations with their parents," and in her experience, positive results are the most common. "I have found that when people get a bit of the truth, they will dig in and try to find more. And, it makes it a better world for all of us when we're on that same playing field of knowing what happened, and how we go forward to heal those pains." Again, she emphasizes the community-building effects of truth.

"In 2015, when we were dealing with the Junipero Serra canonization, many of us across the country said, 'We just have to tell the truth, no matter what.' When somebody like this becomes a saint, the truth has to be told.

We made a pact with each other during that time that really helped, so that it's not only a few people telling the truth, but everyone is telling the truth about what happened, at least on our lands." She pauses then smiles. "I love it."

Talking about history quickly circles Corrina back to talking about the land, because the desire for land is what drove the waves of settlers, and the proper use of both truth and land have been buried here until very recently. In this light, she details how it works for the people today.

"Young people come to our Lisjan arbor, and they help us take care of it. We also have land in Albany held by UC Berkeley, and we help take care of that land, inviting people to come out and help tend it. And, we're hoping to open up another little parcel to community where we have a long-term lease, when COVID is done. It's important for people to have connection. When we lost connection to the land, we lost a part of who we are."

She smiles warmly and recalls Robin Wall Kimmerer's book, *Braiding Sweetgrass*, which recounts that after only minutes on the land, people start humming or singing. "It's our human connection to our Mother, and that's how we're supposed to be." Describing the many young people who come to the land, as well as those who have studied the history, she says, "I see a groundedness in them. They're secure in their knowing.

"I'm hoping that people get on board with this sooner rather than later, because I have a huge investment in my grandchildren. And that means *all* of our grandchildren, so we'll be able to have the tools that we need to fix what we've broken. It's so that we all can move forward in a different way on this Earth."

We sit with Corrina's message in quiet, then thank her for her contribution to the book, but Dahr has a concern he'd like her to talk about. Many people are saying we have to save the entire planet immediately through geoengineering, so he asks for her thoughts.

"You're going to lose a lot of sleep, and you're going to cause yourself a lot of stress if you think that you're going to fix everything. We want to be in that place of fixing it all, but we need to remember to become interdependent with each other again, that we have to live in reciprocity with each other and this Earth again.

"And what does that mean? That means doing it at home. That means learning how to take care of plants so that they can take care of you. It means having a relationship with even the smallest little beings. Human beings think that they rule the Earth and forget things like that with bees and pollinators: if we don't have them we're all going somewhere else. We need them in order to survive. That means planting flowers and making your life beautiful. It means planting food enough to share with your neighbor, and learning those relationships with neighbors again. We forgot that skill of knowing who lives next door and checking up on each other, making sure that everybody has enough. It means creating community gardens and sharing food with people.

"Doing it at home means doing all of those things we forgot when we became human in a different kind of a way, and it's not been that many generations of forgetting. It's about beginning to think of ourselves as *being* that nature, that it's not someplace that you have to travel to. Nature is the tree you see on your sidewalk in the urban area.

"And yes, I hope we have some amazing people of genius who can bring us the technology that we need to clean up the waters and the oceans and the seas and the rivers and the lakes. And I hope that we can all say, 'This is a good idea.'

"I hope that we can understand that Salmon need to run up their rivers, that they clean our waters, and that they help to give nitrogen to the plants so that we can breathe, and that they give us food to eat, and that they need to have the same right to those rivers that we do. And I hope that we open up the dams so that Salmon can live in the rightful way that they're supposed to live, and that we don't dam up places that shouldn't be dammed. Our delta should be clean, and it shouldn't have solvents in it. I hope we can understand that we should be able to do the work that we need to do in our own environments around us to make sure all this happens.

"And, we start in our backyards. It's the old ways of being with all our neighbors, just hands up helping each other out and being interdependent.

And it's awesome! If we focus on that, and on the kindness of people, we can get a lot done."

Living in her deepest recognition, caring for the children, and doing the daily work with courage and clarity; this is Corrina's message, and her contribution.

18

Natalie Diaz (*Mojave/Akimel O'odham*)
The Capacity of Language

COMPOSED BY DAHR JAMAIL

I am of consequence to, or for, or of. Suddenly I am of what has come before me, and I am also what is yet to happen. So how do I try to create a condition in which I am of consequence? Poetry is the way that I've been able to do that.

—Natalie Diaz

I first met Natalie Diaz during the fall of 2015 when we were both in a writing residency in the high, arid desert of far west Texas. During that time in Marfa, Natalie was frenetically busy, as her remarkable book of searing poems, *When My Brother Was an Aztec*, had won an American Book Award, and she was already working on material that would be in her second book, *Postcolonial Love Poem*, which was published in 2020. That book went on to become a finalist for the National Book Award and the Forward Prize in Poetry, and won the 2021 Pulitzer Prize in Poetry.

A Mojave and enrolled member of the Gila River Indian Tribe, Natalie was born and raised in the Fort Mojave Indian Village in Needles, California, on the banks of the Colorado River, a river which she is deeply connected to. While sharing meals throughout our residency discussing literature, politics, and love, Natalie's precision and passion around language was immediately apparent. As an accomplished athlete and former professional basketball player, she told me then she believed in the physical power of language, how words have, literally, a physical energy. The fact that she was a linguist before she became a poet underscores the visceral weight she places upon each word she chooses to use, whether it be in a personal conversation or in one of her incredible books.

In addition to teaching at Arizona State University and writing poetry, Natalie actively works to preserve the Mojave language with its last remaining speakers. For Natalie, the stories of her elders are more important than her own writing.

Hence, it is no surprise she went on to win a MacArthur "Genius" Grant, and has received fellowships from the Lannan Literary Foundation, the Native Arts Council Foundation, and Princeton University. Natalie is a member of the board of trustees for the United States Artists, where she is an alumnus of the Ford Fellowship. She is the director of the Center for Imagination in the Borderlands and the Maxine and Jonathan Marshall Chair in Modern and Contemporary Poetry at Arizona State University, where she teaches in the MFA program. In 2021 Natalie was the youngest poet ever elected a chancellor of the Academy of American Poets.

We spoke with her while she was working back home on the Fort Mojave Reservation in Arizona, and began by asking Natalie to talk about grief, since it is one of the themes of her writing.

"I think a lot about the English language, and I think it exists in a state of emergency," Natalie replied. "What I mean by that is that within the English language, many of us are forced to be in a state of emergency. Sometimes that emergency is visibility because we need to be seen. Sometimes there's a difference between presence and visibility. And in this particular moment, we need to grieve, and the English language and lexicons we have been given for grief teach us that it's not natural to grieve."

She paused, looked out a window for a moment, then continued.

"I think we are seeing the world itself grieve right now. I know my river grieves. I know it grieves after a flood or a storm, or when something wrong has been done to it and it's trying to clean itself. For example, our river, the Colorado, which is the river my tribe is named for and also where I grew up at Fort Mojave, is around 1,500 miles long. It has nineteen dams along it. A lot of invasive fish were restocked in most areas for sport fishing, destroying the natural habitat. They try to build hatcheries, to return some of our traditional or native species fish; however, the river is too deep and cold now, because it's been channeled and dammed. So the few times they've tried to put those fish from the hatchery back into our river, it's like the river just grieves their absence and turns them belly up on the surface. That's also a kind of grief because our world and our practices of living have moved so far away from literal life, which is of course very much connected to the land and to the water."

Natalie went on to discuss how this disconnect arises from how we've been taught that life is about what we accomplish in a day, that *that* is what living is.

"These disconnects are the chasms we have to cross, and grief is one of those chasms. It's something I try to find language for because in my culture we have very intensive processes of grief. They're very ritualistic and intentional. Sometimes we think of a ritual as something that's only done at a certain time, meaning that after the ritual is done, you can live your life however you want. Or sometimes we'll misinterpret our practices as being rote, that we're just doing them, going through motions. But we don't take grief for granted and have a very intentional way with it.

"For example, we don't do it as often now, but there was a time when you never again spoke the actual name of the person we had lost. Which is an intentional way of caring for that person. You speak *around* their name. Rather than say their name, you would say what they had done, or you would talk about a travel they had made, or something that they had accomplished, or something that had happened to them, or even how their family was still here. I think of that as being an essential practice of grief."

She said her poem "Grief Work" is very much about those practices.

Whether it is the climate or the state of emergency of the English language, Natalie thinks of these as "dislocations." Because of this, the word migration has become extremely important to her.

"The word migration has been weaponized, the English language itself is a weapon, but I have learned to reorganize what we normally consider dislocation. Both words tell of a relationship about choice, about natural conditions, and I want to imagine them both as having the possibility of return. Migration is important to how I think about grief. Migration is a very natural movement of animals, of seeds, and people. Grief is similar to that relationship of homeland. It's not something you ever leave, but it's something that becomes a part of you, and it becomes a part of everything else you do, wherever you are. Grief lives in a strange place." She paused to collect her thoughts, then continued.

"It's an out of time place. Natives are timeless as a natural condition of the worlds we've lived, especially because America happened *after* us in many ways. So there's a way that, not only in our spiritual beliefs, but in our very *now* American lives, we're also out of time, working the way grief works. It's happened, but it's always still happening, and it's also shaping what I do next."

Natalie went on to say that this is one of the ways she "refuses to be prophesied" by the English language and works hard to be capacious in English.

"I want to imagine my own lexicon. So the lexicon of the body is paramount. I must find a language and an imagination that allows me as a Native woman, as a queer woman, as a Latina woman, to have a body. Sometimes that body is in grief. My body is its own lexicon and I also fight for a language, in Mojave and English, that helps me to hold it in the space of love. And I don't mean love as a blanket definition. It's as violent as any human thing, and also a storm or a river."

Natalie's words reminded me of the impacts of unacknowledged grief that wash back and forth across American society. I asked her if this was just a continuance of the original and ongoing denial of the genocide of the Indigenous peoples of Turtle Island, and how the unacknowledged grief from that across both sides, the perpetrators and the victims, destabilizes everything.

"I feel extremely lucky to be a poet, even though I don't think poetry should be a luxury. Audre Lorde said poetry is not a luxury, yet it *was* in America. It is *lucky* in a Mojave sense. It is a gift that I've been given, and it offers me a different sensuality in language that allows me to articulate questions. One of the questions poetry helps me to think about is the difference between implication and participation. That's a little bit like the relationship you brought up with panic versus urgency."

Natalie paused, after mentioning that she was thinking more about grieving. She went on to share how her father was in Vietnam, and still suffers from detrimental injuries. It was two days from coming home, in a supposedly safe area away from the fighting, and he was hit by a missile, leaving him with wounds that continue to bring him daily pain.

But upon his return to the United States, Natalie's grandmother was told he had been killed. Her father ended up in a hospital in San Francisco for several months.

"She [his mother] had no idea he was alive, until he hiked home and that's how he made it back. And that's when they knew. I believe there is a difference between emotionality and sensuality. I think sensuality is not the same thing as senses. The English language gives us these emotions or senses, like it's eager to identify five senses, but it fails to offer us sensuality. Part of sensuality is unknowing, and in America you must always know. Knowledge is such a currency that I feel like unknowing is an unfortunately lost sensuality. How could we not be lost in what is this life, not just life in the human body, but life in and of the world? In some ways to lose someone is to lose a part of ourselves, and then to realize that what is left of us, who we still are, is part of something else. Grief is a sensuality that allows us to break, and I understand the importance of breaking. I see in my family what happens when you don't break, when you are taught you must endure, absorb, carry on, fight through it. In the poems of my brother I explore how he was taught he couldn't break and now he's broken. Yet I don't think broken is bad. My brother is broken but he is not bad."

In 2013, for the first time in her life, Natalie broke.

"I reached a point where I was on my mother's couch for two to three months, and I couldn't leave it. I don't know what happened. I'd been doing so much for so long, doing my language work and being on the road and having a book and having been an athlete. It was a part of me my family had never seen before. Nobody really knew what to do. All my family lives here on the rez and I remember my brothers stopping by to check on me. Once I overheard them trying to have a whisper conversation. They were like, 'What's wrong with her? Why doesn't she just get up?'"

Natalie did not know what was happening to her. She does not cry often, because it is not the way she learned to be emotional. Rather, she has often shifted her emotions into physicality. She became an athlete, and that was how she dealt with her feelings. This emotional physicality and way of touch is her central relationship to the world.

"My brothers didn't have that, that generous space basketball offered me. They played basketball and taught me basketball, but for me the game was a way to reorganize my hurt or pain or unknowing. Basketball didn't give that space to my brothers. One of the reasons why I think grief is important is because there's something we've been taught about durability and duration that feels unnatural to me. We watch the Earth break, we watch animals in certain populations break and respond and react and reorganize. Our Mojave stories allow for that breaking. I think part of that breaking is related to migration, to the ways we shift and change and continue to become and unbecome human."

Natalie said there is something about duration to her that feels like a numbing of the body.

"Duration is dangerous because if we believe in it too much it can make some of us think we deserve more pain. There's a numbing of the senses. As we begin to think about duration, we also become less intentional, even in ways we talk about climate or the environment; for example, the idea of sustainability is so wild to me. That anyone would ever imagine everything can be or should be sustainable or sustained, is incredible. It's like a love affair with stasis, isn't it?"

She pointed to where we each live, and our sensual relationships to the land.

"Those relationships are really what make us alive, otherwise the body just becomes a series of gestures. That's what feels difficult, that we're so caught in these production cycles, and they are not very sensual cycles. It's heartbreaking. Personally, that's what I did. My father didn't know he had PTSD when he came back in 1965 because it wasn't even a concept yet at that time. So you either sucked it up or you were bat-shit crazy. Those were the two alternatives. Even how we talk about violence. I think a lot about that word, and I know you both know that word very differently [Stan and I both have experience with war]. I think about the structures violence places us in and the things it normalizes, again such as the need to endure and that idea that you can't break. How can war not break you? How could it not, given the fact that it is designed to do just that, to break you and them?"

Natalie mentioned Japanese American author Brandon Shimoda's book *The Grave on the Wall*, given he is questioning the idea of monuments. I'd mentioned Veterans Day, and how the dominant culture urges us to thank veterans for their service, and the hollowness of it all. Natalie said the problem is evidenced even in how the dominant culture marks what we have done abroad, which is itself a refusal to acknowledge what we have done.

"I like the idea of how we're tearing down monuments, but we are also putting up new ones in their place. The monument, in some ways, disrupts that relationship with the land.

"I wonder a lot about these gestures toward and against monuments, about the uselessness of the monument and how the 'new' monument believes it's providing a healing of what was once a marker of a victorious violence. There are direct unarticulated links between this reified insistence upon endurance, the sucking it up, the denial of genocide or wars or the true violence done as the cause of the climate crisis."

Natalie shared that she is unsure if she's lived long enough to fully understand what it means to be of the land, but added, "I know my stories and I know that I know the closeness. I know that I feel closer now being home, and coming to a more mature relationship with my river. It was a place of play for me for a long time. As I get older, and it becomes more endangered, I have a different relationship with it."

Stan and I have both discussed in our own ways the importance of doing what we do for the Earth, regardless of any possible outcome, no matter what. Given how overwhelming the current situation on the planet is, this has felt like the only way to comport oneself properly.

"I was really struck by your phrase, 'no matter the outcome,'" Natalie responds. "I am struck by the insistence that these things are a part of the state of the climate right now. I don't believe this is us having control over the land. This is the land telling us it's tired. It's tired of sustaining us when we can't reciprocate the relationship that I believe was set out for us. I don't have the articulation yet, but I do think there's something about when we do acknowledge what we've done, or even our refusal to acknowledge what we've done. It's still such a dominion, right? Like we are holding dominion over the land and we will direct it. We are definitely a part of the momentum of what's happening to water, to Earth, to climate. Yet this is also very much the Earth telling us it's exhausted. It's ready to start cleaning itself. It's ready to move on to its next iteration."

Natalie shared with us that her father is Mexican and Spanish, not Native, yet grew up on the rez alongside her mother's family. Thus, her family has all had a long-standing deep connection to the desert.

"We have a big volcano cone, Amboy Crater, so at some point everything here has burned up. So fossils are not as abundant. We have the rift zone and the fault line right up the road, and lava fields and huge salt flats. Along with all of this we have the river and we're in a valley. I've always known the Earth is on a track of its own, already set in motion to do what it does. My father always spoke to this, and he is still very interested in earthquakes. My father created in me this way of looking at the Earth, that we are only one movement in its great momentum. But now we have the Anthropocene, right? Which is problematic because it re-centers humans, as in it's still *our* time, a time that's just about *us*."

Natalie recalled when we were in our writing residency, she, I, and Spanish translator Will Vanderhyden.

"I had the luck of being alongside Dahr and Will," she said, speaking directly to Stan. "In the evenings we gathered sometimes and had conversations. We called ourselves 'Team Doom' because we always arrived at talking about climate disruption or all of these different crises in the world. One night we were talking about if people should keep having children, or should I be driving a Prius. But one of the things Dahr said to me was this: 'Can I be honest with you? This has been terrifying to know these facts about the climate.' But Dahr went on to tell me that he was just trying to reduce his carbon footprint as much as possible because he was thinking of the generations to come after us. That has stuck with me and really changed me. I even remember the next morning waking up and being like, the world is new, and I'm new in it."

This was the first time I'd come toward an iteration of that "no matter the outcome," I must find a way to comport myself in the best way possible amidst the catastrophes besetting Earth. Although it wasn't until I met Stan, years later, that I literally came to terms with this concept.

"But this is something that I'm really grappling with," Natalie continued. "I feel like we don't really have a language to talk about how small we are and how we might cause damage that on the human scale is immense and overwhelming. It doesn't mean we won't damage other life-forms, yet I think that life is so much larger. It's hard for me to find the practice of that except in trying to think toward it. We are very much impacting and destroying. We are a destructive species, yet life is also entirely larger than we are. So I feel like the scope actually shifts, so it's not so much about us. What are we going to save? What are we going to prevent? We're now dealing with tiny numbers and degrees. Whether it's talking about temperature or the ocean rising, there will be life after us. So when I originally heard this comment, 'no matter the outcome,' to me, that might seem like the most essential comment. 'No matter the outcome.' Who do I want to be as a relational being or person, no matter the outcome? I don't think I even have an answer now, except that that feels like the question I needed."

Stan talked about carrying this question of what is ours to do, no matter the outcome, against his personal backdrop of the moral wounds he sustained while in the military. He mentioned how the VA has the term "moral wounds" for what happens to soldiers who were sent to war believing in the cause, only to see other soldiers commit atrocities. Further, he added that whether his life is shortened by the chemicals he was exposed to (Agent Orange), in light of such wounds, he believes that the manner in which we live our lives is the most crucial thing. "It's about the manner in which I engage every living being that I come in contact with that is the most crucial thing. *That's* the relationship." Dialoguing directly with Natalie, he pointed out the constant change and motion of relationship, the *constant exchange* being what life is about. "I think we get hung up on the outcome and then we make tons of mistakes trying to make that perceived outcome happen when it's only coming from a partial framework, a partial understanding of what's going on."

Natalie began to ask questions while thinking aloud.

"How did we get here? How do we move on and respond? What's getting in the way of that? If there's a quick answer to these, it's that humanity is a small part of this much larger thing, and we must get back to being in relationship to the land and water. What might humanity be if it could be back in relationship? It's the large question of the ways we try to push ourselves through a day. To be who we are. To say because I'm a poet, this can be who I am in a day, or this is how I might treat someone. Or this is what I might always be intentional about or central to in language when I speak to my mother or my lover or my friend or a stranger. So I think those are the small ways. Or how do I visit my water? And then how do I use it in a day? Or how do we work in the garden? Then we have the larger world that is not asking that question, no matter the outcome."

Natalie paused, then said this reminded her of abolitionist work, or people who have been traumatized or suffered the moral wounds from war, and how important it is to give these things language.

"That's why I think it's so impactful that you are working on this book because sometimes a story, in a Western sense, lets us forget this. Indigenous story works differently, because the intimacy of its language is like touch. So, suddenly, like putting the word 'moral' in front of wounds, that's a whole different way of being held, and it's a whole different way of learning to love yourself. Sometimes we forget the simplicity and power of a single word, of a way of language."

Natalie thinks that how we iterate or replicate or commodify things flies in the face of what the land told the human body to become, what that body might offer, that this is what real abundance is.

"It was Creator pulling us up from the land and saying this is who you are in language, but it happened while land was happening. Then to say this is where your water will come from as he moved his lance into the Earth, and the water came out. Then to say, this is how you hunt. Language is one pathway to return us to land the way the Creator gave us language to identify ourselves as both land and water, as being *of* each. Language has a power of return to its first iterations. The first time Earth was said, it was the Earth itself. Indigenous languages hold these knowledges but even the English language can be made more capacious and open to these relationships. In language, we'll always return back to the land if you know how to look for it."

We sat in silence for a moment. Natalie again looked out the window, pondering, then looked back at us and began to share her thoughts. She began to discuss how her elders worry that when young Mojave people dream, they might not know what their dreams mean.

"My elders worry that young people might be dreaming in Mojave, but since they don't speak the language they might not understand the dream. And our gifts are given to us through dreaming." Natalie paused here, thinking, looked out the window again, then continued. "This happens not just in sleeping dreams, but just dreams that happen all the time. They might be arriving to us in Mojave in ways that maybe we just don't remember, or we don't understand."

Natalie has questioned the ability of the English language to express what needs to be expressed.

"What does it mean that it [English] was built? Or it was uttered against people like me or against bodies like mine? It's [language] not the body, but just one of many ways one body carries itself to another. It is physical. That's why, for me, calling language 'touch' helps me remember the physicality of it. You have to be careful with what you say, because once you let it go, it's always moving, always touching things. So there's a great power in that. It's why we can remember the things people have said to us however many years ago, and it still hurts us, or why tomorrow when someone we love calls our name, there we will be brand new again in that moment. There's power in that.

"I used to say I had to break the language, but I don't know if that's necessarily the way I think about it lately. Again, language is not the body, and it's the person who is speaking themselves who has that capacity, that possibility. We don't just speak language but we also enact it. Language is the imagination of how we might hurt or touch or hold someone. We've seen the capacity of language. We've seen all the people who came before us live and love in it, whether we learned love the hardest ways or in the softest ways, or in a mixture. It's lucky to imagine that each of us are who we are in the midst of all of the places we've been, and all of the language we've heard or been told."

Stan, having been deeply moved by reading several of Natalie's poems prior to our interview, asked her if she ever thinks of her work, her art, her language, her poetry as a contribution to the question of how we are best to comport ourselves, no matter the outcome?

"It's a lucky question, and all of these questions have been, and in hearing you as well.

"I'm thinking of Dahr's book *The End of Ice*. When reading in certain moments, when I knew that Dahr was back in that space of the mountain and thinking in particular of how the cold always sticks out to me because I'm from the desert, there was a way that your language moved. Like that wind you described up there. I think of it as the overwhelm, and to me that is essentially what a body is for. It's to be of, and alongside, and within the overwhelm of life.

"The Catholic church has called it 'the ecstatic' for so long. I actually think the ecstatic is not away from the body. It's back into it, before the very danger of what it means to live. Not that you always have to be on the precipice of danger, but I do think danger is always at stake. The body is always at stake. Some of us know this in more terrible ways than others, but I think something that has been really important for me, especially because of some of the ways my work is received, is this phrase I keep returning to. For me, [it is] to write in poetry, especially when I'm allowing myself a full body, a body that's capable of pleasure or love while also being aching, or a body that's capable of possibility in an impossibility. Maybe those two things are the same. But I think a lot about how I bring myself to bear on the English language, which has been brought to bear on my people in my land forever. How do I bring myself to bear, not to be visible, but to be physically present?"

Natalie discussed how this is one of the reasons why she needs language to move faster, and to touch her more than it has in the past, or to touch her differently.

"I've written about my real-life brother, but on the page he becomes 'the brother.' I am not writing about my real-life brother but I'm writing about my emotional experience of that relationship, and I invent a brother on the page to do that. Some of these memories or imaginings are difficult. But there's a way that on the page, in the language of poetry, I can look at him the way I think he deserves to be looked at. He does not deserve to be ignored. He deserves for me to allow him to break. There's also a way that I hold him physically on the page in ways that I can't yet hold him off the page. There are two kinds of love that happen in language that are important to me when I write poetry. I love myself better there on the page than I can love myself during some days. I love my brother better on the page than I can sometimes love him in our life. So to me, that's how I make English able to hold that which becomes extremely physical. I need it to be all of the things it has been, including the violent things, which is why a lot of the work is very etymologically rooted. That also mirrors how I'm trying to make it hold Mojave thinking."

Earlier in our conversation Stan had mentioned how he works to allow the students he teaches their complete autonomy. Natalie recalled this, as she continued.

"I think that's beautiful, thinking of that relationship of your students to have autonomy in relationship with you because that's the way the reservation works or any of our communities. My autonomy only exists in relationship to everyone around me. And so this phrase has been really important, but I keep telling it to myself: that I am of consequence. I wish we could find ways, or that we had this concept in our Western English language, to express that, that I am of consequence. That reminds me of your question, 'No matter the outcome, how do we comport ourselves?' I am of consequence to, or for, or of. Suddenly I am of what has come before me, and I am also what is yet to happen. So how do I try to create a condition in which I am of consequence? Poetry is the way that I've been able to do that."

Having played professional basketball, Natalie thought of how all that she had been discussing connected back to her experiences in that sport.

"You're always of consequence because everyone is moving in space, and space becomes so important. What hasn't happened, but might happen in thirty seconds, or what might happen in ten seconds, or if she moves there, what space opens up and what does that mean of what I see? Basketball happens often in the periphery which is a natural decentering. I still have that sense. I don't know if it's a curse or luck to know what's happening over here or over there. I'm always a little overaware. But related to how to comport myself is to remind myself that I am of consequence and I can be a presence of my own body, even in small ways."

The cover of her first book is of her little brother. The cover of her second book is a photograph of her.

"Part of that was me acknowledging, first and foremost, before any audience, that *we* are the bodies at stake. I think that is important because there's always so much concern about what people will think, or who's in the audience, or who are you writing for? But in some ways I'm writing for me, and for us, and that feels really important. Realizing that I am at stake, and I am of consequence, and sometimes those aren't great feelings to have in the scheme of what I have done, or as I am considering the choices related to what I might do next.

"Not to mean that we shouldn't do this, but relationality is not easy in some ways. I understand why people have dodged it and run from it because to be relational is not easy. I think of Dahr, with your work, knowing all the things that you've known to have that physical relationship with the Earth and with these places. Then what we've separated. You returned to it in a nonseparated way, but we often separate the intellectual from the physical; to know the things you know, of what is happening to those places, and to have that relationship like that is a burden in some ways. I don't think burdens are necessarily bad, but that's *something*. Your body is not like my body because of those relationships; what it has borne and what it bears is different than mine."

Deeply moved by all that Natalie was sharing, it occurred to me how love, connection, and obligation are similar to the linguistic medicine of which she spoke. I was reminded of our sacred obligations of serving future generations and serving Earth, and how when I realized the linguistic connection of this, just having that articulated to me changed my entire ontology. I shared this with Natalie and Stan, this connection being made for me through our discussion, and through Natalie's words. I used to feel so sure life on Earth is over because of the climate crisis, but I shared with them how when I was reminded of our obligations, I fell in love with this planet more than I ever have, and feel more connection now than ever before. And that happened *because* of the change of language.

"This conversation feels so possible," Natalie said. She remembered how Stan opened our time together with a prayer in several Native languages. "Even how Stan began the conversation with our Indigenous languages, how our Native languages change everything. Even for people who don't understand them. I didn't necessarily know what you were saying, but meaning is the least important thing. I feel like you can relate and connect and touch without meaning. Meaning is context. These other things are conditions and they're conditions in which we have all arrived at life or the people before us. That, for me, really shifts it. It's interesting because we're in different spaces, so we might say the physicality of this is not necessarily happening in the same way if we were in a space together, yet that language for me signals the shift. Like I know now I am not just listening, but I am attentive in a different way. So suddenly my body is alive in a different way. How do we comport ourselves?"

After thinking for a moment, Natalie mentioned the Mojave way of gathering, Matakyev, because as she said, it felt "essential" to her. She wondered what it has meant to be unable to touch during the global pandemic, but also considered how we were only considering one kind of touch.

"I think there are certain things we can't deny about physicality and physical touch, but there was a time when Mojave were so spread out over the valley that they came together either for a death, or once a year at these large gatherings. So Matakyev means to gather the bodies from across the land. So even when we were apart over those great great temporal and spatial distances, we were still in touch and connected because we were of the land and living. I think it's a return to language. Like moral wound isn't a backflip, it's putting two things together that should have always been together. So just imagine had you been given that earlier, Stan; just imagine who now will be changed because of that."

Natalie paused then shared how this is one of the reasons why she's so interested in how Indigenous languages are closer to the land connection.

"They can change a space or keep us at least aimed toward that final question which now feels like a new beginning question, 'no matter the outcome.' Because I feel like that's what our languages were built in, 'no matter the outcome.' When we were made, I feel like we were an imagination of the land, and they had no idea what we would do. And of course we've screwed up many times, but we were made and they had no idea what we might do. So there's something of that that is also old, that 'no matter the outcome,' like as a plant grows, it has no idea what's going to happen to it, yet there it goes, it just pops up."

She looked out the window, then back at us, and continued.

"We have a garden in the middle of this desert, and it's amazing, that almost unknowingly these things are willing to grow for us. It feels like pre-preknowledge. They grow because that's what they were born to do, to desire life. This conversation feels really lucky. I was able to arrive here not needing to know anything. These have been rough days with this, but now I feel like, 'Oh, there's a whole day ahead of me, like a really long time. . . . Alright, what comes next?' I'm gonna fall right in to whatever comes next, because of that thought, 'no matter the outcome.' I feel like I'm different now. The relationship you all are building, what you're gathering with this book, I think it's going to be so important for anyone who manages to arrive at it."

Natalie mentioned how essential questions like this are for younger Indigenous people and their relationship to land. "I feel like arming people with that final question, because it reorganizes you, like now I'm completely rearranged going into the day, which feels really lucky," she said.

"I think back to when I was doing early research when I first started my language work with my elders. The tribe had mostly hired me to go through archives. So I was looking at all of these old notes from the mid-1800s, and there's this guy named George Devereaux, who is a real jerk, one of these German and French philosophers who came and did their psychology-of-sexuality studies on the natives."

Natalie brought that up because of the institution of the idea of blood quantum as a settler colonialist tactic of pulling Mojave and other tribes away from the land. It was simply another way they worked to take the land for themselves, even though they had no relationship to it. Of course it also served to remove the Indigenous from their land, in hopes that they would just die off, or become assimilated, eroded, erased.

We were approaching the end of our time together. After a pause, Natalie thought of one more thing she wanted to say.

"I guess the last thing I'll say is that one thing that's also important lately is the language I use in my home, even though the language I'm most known for is on the page. My partner is Black. Her family is from a long line of farmers in Mariana, Florida, and we talk a lot about land and Indigeneity. It's not about blood and it never has been. It's about how you arrive and build a relationship with the land you're on. And then once you're there, it's also about how you receive others. That's the reciprocity between arriving and receiving, as well as the practices by which you tend to your relationship with the land. That was such an efficient weapon that they used, that the Native peoples should be denied believing they are Native, or of their tribe, because of all the ridiculous things, because of no reservation, as though suddenly that became the ideal."

She thought some more, then continued.

"Here, we know our reservation boundaries, but because we're in an area that is not heavily populated, we have access to everything. There are state parks and national monuments, and wilderness areas here, but nobody is really out there. So I can still go to my creation mountain, even though it is located in a different state. I've done some language work with some folks out in California, when the Mojave actually traveled extensively up and down the coast. There were some Pomo friends of mine. We had met at the national anthropological archives for this Breath of Life festival. All of us were looking through old archives and I had just given a presentation about Mojave runners, how we ran to the coast and we carried language. We have these Mojave words dotted all the way up the coast. We traveled out of curiosity, and we followed animals and birds."

When Natalie had mentioned the Mojave way of gathering, I had not been able to discern the word, so I later communicated with her to ask her for the spelling of it.

She said the word again, and its spelling, Matakyev, and added, "'Amat' is land, 'limat' is person, body."

I asked her if the first three letters of Matakyev meant "land."

"Land or body," she replied. "Mat- is the prefix that holds both. Context determines which. For us, they are tethered, land and body."

19

Melina Laboucan-Massimo
(*Lubicon Cree*)
Paradigm Shift

COMPOSED BY DAHR JAMAIL

We need a paradigm shift of understanding that we aren't in this hierarchy with Mother Earth, that we are in the sacred hoop. That's something that still needs to shift. It's how we right our relationship with Mother Earth.

—Melina Laboucan-Massimo

When she was only seven years old, Melina remembers her family actively blocking the road to their village in resistance to resource extraction companies aiming to desecrate their homeland in northern Alberta. This, after her father had to be hidden by his parents from government officials who continually came by looking to take him to a residential school, where beatings and other atrocities against First Nations people were part of the curriculum.

Melina remembers growing up on the sacred Earth, drinking fresh water from streams, breathing clean air, eating straight from the land. When we spoke with her, she was thirty-nine years old, and told us she can no longer do that in the land where she grew up. In just a single generation, giant oil companies and their tar sands operations have toxified the air, water, animals, and vegetation in her traditional territory.

Those experiences having set the stage, Melina has responded by devoting her life to working on social, environmental, and climate justice issues for the last two decades. She is a fellow at the David Suzuki Foundation, the just transition director at Indigenous Climate Action, and the founder of Sacred Earth Solar. Her work has included being the climate and energy campaigner for Greenpeace Canada, and working with the Indigenous Environmental Network internationally. Her studies, work, and campaigning have taken her far afield, including Brazil, Australia, Mexico, and across Canada, where she has shone her light upon Indigenous rights and responsibilities, resource extraction, and climate crisis impacts.

Tragically, Melina lost her sister Bella in a suspicious and tragic death, whose case remains unsolved to this day. This led to her becoming involved with Murdered and Missing Indigenous Women in Canada, because, as she would tell us, "Violence against women is violence against the land."

Melina is currently the host of a new TV series, *Power to the People*, which focuses on food security, renewable energy, and ecohousing projects for Indigenous people across Canada, in addition to serving on the boards of the NDN Collective and Seeding Sovereignty.

We caught up with Melina while she was in a studio, having just completed a long podcast discussing environmental issues. We began by asking her if she would discuss her ideas of what our obligations are, to serving the planet and future generations of all species.

"I feel like I try to explain this to non-Indigenous people, and sometimes they're looking at me like I'm speaking Chinese or something," she said, with a wry smile, wasting no time in jumping straight into the depths of her point. "I feel like I'm speaking a foreign language to them, because it is really about a paradigm shift."

Melina feels this way because she believes, as she put it, "The climate crisis is the direct result of the Western capitalist industrial complex, coupled with patriarchy and all the other 'isms' that exist which separate humans from Mother Earth." Hence, she sees the solution can be found only in "working to reestablish our relationship with Mother Earth, reestablishing our treaty with Her, and our treaty with the four-legged ones, the winged ones, the fish nations, and all the different beings that are not just in human form."

Cognizant of how deeply her life is impacted by the overuse of technology, she held up her smartphone to underscore her point. "This technology brings us out of our bodies and into these screens, and out of our immediate moment with Mother Earth. Then we become very cognitive beings, and we don't need any more cognitive beings."

Her point was not that we don't need to think, but that she sees this overcognition as one of the causes of capitalism and White supremacy, along with how Western society in general functions. "This has been very destructive because it creates a sense of hierarchy, as if humans are somehow on the top of this pyramid, which, as Indigenous peoples, we know is completely out of alignment with the understandings of natural law and the sacred hoop, and the ways in which we engage with other living beings, other humans, and Mother Earth."

For Melina, the impacts of colonization, both physically and mentally, destroy an inherent balance crucial to the health of all. "I can even feel that within myself sometimes, where I feel like I'm becoming disembodied. That is because everything's here [she pointed to her head], up in cognition and not in heart-based reality."

She acknowledged how, after having been campaigning for so long, she'd had to exist in the cognitive state intensely, including testifying before Congress and communicating in diplomatic ways. "So I'm presenting information in a way that is 'unemotional' or not heart connected because I need to exist in this cognitive way."

Melina believes we will, literally, have to find the ways in which we will continue to exist due to the climate crisis. The first step in the new direction entails getting out of our heads, in things like learning to garden again, hunting, fishing, trapping, when possible in zones that have not yet been made toxic by the industrial growth society.

She paused briefly, then put it this way: "Returning to Indigenous ways of knowing and being."

Melina remembers growing up and being in the horse and wagon with her grandparents. She was raised within those older teachings and ways of being, and watched them in her life. She is keenly aware of how this is not necessarily the experience of a lot of Indigenous peoples due to the impacts of colonialism, and the forced removal of people from their homelands.

She acknowledges how much her understandings emanate from her coming from a land-based community that was not forcibly removed, "other than my parents' generation being forced into residential schools and day schools, which impacted the social fabric of our people," and she feels fortunate. "As a younger campaigner, I was blessed or privileged with that understanding of being a land-based young person, a five-year-old on the land and seeing the beauty of Mother Earth in her 'natural state.'"

While logging and other impacts were already ongoing, Melina noted how a "pristine cleanliness" remained. Again, she could still drink from streams, a vital thing to consider. "That's very different now. I'll be turning forty this year and you can no longer do that. You cannot be in our traditional territories and our homelands and drink the water. A lot of times you can't breathe clean air; there's fragmentation of the forest and the lungs of Mother Earth. So the impacts are very real where I come from."

She has what she describes as "pre-immense resource extraction memories" which she can compare things to. This informs her perspective, her thinking and feeling.

"So eventually we have to come back to the full circle and I don't know what that's going to look like with climate change, because in ten, twenty, or thirty years it's hard to know what's going to happen. That's why I've spent the past twenty years trying to do what I can, which is like a little drip in the pool of all the different actions that people have been doing to protect Mother Earth. But it's hard. It's really hard to know what's going to happen. So I feel like trying to reestablish relationship with Mother Earth is the number one priority, but there are a lot of distractions."

Ironically, one of the distractions has been activism. She sees herself and the younger generations taken offtrack by technology at times, and this is why she feels she must dig deeper to remain focused on what is paramount. She talked about what it was like to always be "going" and "doing," often at a frenetic and relentless pace. Therefore, she was actively working to try to step off the treadmill of activism, because while having good intentions, she had made herself sick.

This led our conversation toward asking Melina if she would share her experience of having her people's land desecrated while they were still living on it, something most non-Indigenous activists have never experienced. Before answering, Melina acknowledged it would be difficult to talk about that without becoming emotional.

"When you come from an impacted zone like the tar sands, it's like a direct violation to your spirit," she began. "It's that I can't breathe. My eyes are burning. It's that oil spills, for us our biggest ones, are at our home. It's the toxic burden. Our community was blockading in 1988, when I was seven. Our family community, my grandparents, my parents were all on the front lines, and I was too."

Melina said she barely has memories of it, and wonders if she blocked them out. "I barely even have memories of it because I think it was so traumatic. You are this little kid and you feel that sense of conflict. And I knew what happened with my dad and my aunties around residential schools, and them being taken and stolen from our community, and the trauma around that. So we knew intrinsically that we were never truly safe."

It was that trauma, coupled with the experience of being born onto a blockade, which Melina feels she inherited. People ask her what made her decide to do the work she has done and continues to do. To this, she said, "I don't really feel like I ever had a choice." Melina was born into defending the land. Also following the call of necessity, her mother became a psychologist who now works with residential school survivors.

Her family moved to the city, which was also a traumatic experience, given their traditional community had no running water and only dirt roads. To find swimming pools, libraries, running water, and resources abounding that did not exist where they had lived was a shock. But what was even more of a shock was the realization that the resources being utilized were being taken from their home territories, and none of them were going back into those communities. "That really politicized me from a young age, during my teenage years. I started working, and I finished high school when I was sixteen, then went into university immediately and started doing community organizing as a youth worker, and tutoring inner-city Indigenous youth." She was nineteen at that time, and hasn't stopped working for Indigenous communities since.

"I saw the discrepancy, and knew there was a problem. The fact that Indigenous communities are being treated this way made me want to seek justice and bring that into the awareness of Canadians. This was before the Truth and Reconciliation Commission. No one even really knew about residential schools. It wasn't in the history books. So there was a lot of having to combat a lot of ignorance and a lot of people that didn't know the history of their own country."

Melina's pursuits for justice for her people have been along all the paths where the needs exist: Indigenous food security, renewable energy, and affordable housing for Indigenous communities, wherein she has been working at a blistering pace for twenty years straight.

One of her projects with Sacred Earth Solar is providing mobile solar units for the Tiny House Warriors, who are on the front lines fighting to stop the Trans Mountain expansion pipeline. They use the mobile units to charge their cell phones in order to be able to record and document militarized violence forced upon them by the Canadian government, along with documenting the construction happening across their homeland. This work also focuses on stopping so-called "man camps." Studies have shown these industrial camps are directly associated with increased rates of sexual assault and violence against Indigenous women.

In 2016 Amnesty International stated, "Indigenous women and girls suffer the highest rates of violence in Canada."

These facts underscore what Melina told us.

"Protecting our land, and protecting our women, go hand in hand."

Melina's story underscores what is ubiquitous across Indigenous communities: that the climate crisis cannot just be put in one category, because *everything* is impacted. "We can't exist in these silos anymore. It's not enough to have a climate justice or environmental justice silo, and then an Indigenous rights silo. They're all connected, and that's what we've been telling people for fifteen years. You get a little frustrated trying to communicate that to people."

She shared how she has been in conversation with the leadership at Standing Rock while working on the television series she hosts, *Power to the People*. She was reminded of the importance of not thinking in boxes, of the need to remember the sacred hoop, and how it is all encompassing. "It's not just one way of seeing things and one way of being, or one way of doing. *Everything* is important. The climate crisis is the issue of our day, of whether or not we survive on this planet, but we also need every single person to be involved because everyone's impacted. We need a paradigm shift of understanding that we aren't in this hierarchy with Mother Earth, that we are in the sacred hoop. That's something that still needs to shift. It's how we right our relationship with Mother Earth."

In the effort to stop the extraction of trees and oil in their territories when she was a child, Melina's family and community had to blockade the very road that they depended upon for their own supplies. This level of commitment and protection came from the knowledge that they were defending their way of life. As long as the land was protected, their ability to feed, water, and care for themselves was intact. Without that, they would be forced into the cities, and forced to rely upon the system that was destroying their territories. Melina referred to this as the "dependency state."

"The dependency state was fully flourishing by the 1980s in Canada. The colonial policies in Canada were severe. We didn't have the right to vote until 1960. Indigenous peoples couldn't obtain a lawyer for political purposes until at least 1960. We couldn't gather in groups of more than three until 1960. This was the blueprint for the apartheid system that was implemented in Canada, then later in South Africa. The Canadian colonial state developed this blueprint of how they treated and undermined Indigenous peoples, to pull apart the social fabric of our communities."

For a long time, her family managed to avoid this. Helped along by living so far out in the bush, and the fact they were very self-sufficient, they hadn't signed any treaties with the Canadian government. They were remote enough that treaty surveyors did not find her family. Melina's grandparents only spoke Cree, and never went to residential schools. But as time went on, she had aunts and uncles who started being taken from their communities, when Indian agents working for the residential schools began finding them. Her grandparents managed to keep her father hidden from the agents, so he was never taken and forced into residential school.

"He was the youngest, and they realized how horrible the residential schools were from my aunts and uncles, his brothers and sisters. They hid him until he was nine or ten. They managed to have him only go to a day school, so he didn't have to sleep there, because obviously we know how horrible sleeping at those schools was for kids." Melina paused for a moment, given the weight of what she was sharing with us. I had to remind myself to take a deep breath, while taking off my glasses to wipe away a tear.

"That was the kind of sadness that was instilled in most Indigenous communities where the kids were stolen. There was a deep sadness that I felt genetically. There are ancestral wounds, it's in our DNA. My dad became an alcoholic for a good portion of my life because he was so sad. He had such immense loneliness because he basically grew up alone, just with elders. That's how he learned a lot of stories and history and language. He didn't speak English until he was in the day school, and then that's when they started beating him. I still get sadness in the fall because that is when all of the kids were stolen. It's very traumatizing and triggering, and this wasn't in the history books, so no one knew what happened to our families."

Melina also grew up with a fear that at some point she too might have to go to a residential school. "There's this immense sadness that I felt around how when all the kids were stolen and taken, and I was by myself, that I eventually would have to go into these horrible schools. And so of course people are going to have these triggered and traumatized ways of being, which are really hard to undo if there're no support systems for healing in our communities. We have ceremony, which is important, but sometimes the people leading the ceremonies are also traumatized and triggered themselves. There's a lot of healing to do all around."

Melina is acutely aware of how the work she does is in and of itself both triggering and traumatizing. She discussed the challenges she faces while trying to communicate with White environmentalists sitting in offices in places like Toronto or Washington, given that she comes from a community living in a state of chronic crisis while battling multinational corporations already active in her territory.

"It's a very triggering experience to live when you're a campaigner; because you're constantly engaging with adversarial forces, you become very combative. I feel like there're two ways of Melina, one of which is the Cree side where I'm quieter and more introspective and want to hear other people talk first. The other is the Melina that has had to campaign within White-dominated spaces, and I need to get my word in edgewise and be very loud and vocal, and as smart and articulate as I can sound. Those are two very different people culturally. So to be a campaigner in a White-dominated space is very triggering because you have to speak about trauma in a way that White people will understand, but they don't fully understand. Then it's traumatizing to talk about because you're talking about all of these historical and/or present-day realities of colonial policies that exist, or previously existed, that are racist, like the treatment of [the organization] Murdered and Missing Indigenous Women, and all of the other things I've worked on. People who have not experienced these things can only understand to a certain extent."

Melina has learned not to talk about these things as much as she used to, having given presentations for years about all of these issues. Each time she did so, she was aware of how significantly impacted her adrenal glands were. "I learned I was existing in life for the benefit of other people's education, as opposed to existing for my own healing. That healing wasn't happening because I was constantly retriggering the trauma."

Melina paused there, took a deep breath, slowed her speaking, and continued.

"I hit a point where I collapsed. My body just collapsed, and it was shocking for me, although no one around me was surprised. I was shocked, because I'm an athlete and able-bodied, but my body was giving me all these signs. But that is what I'm talking about, this disembodiment that we have in Western society that tells us we exist only in the brain. We don't listen to Mother Earth, and what our body, which is so connected to Mother Earth, is saying."

Melina explained that she felt like it was enough for her just to do the work, even if it meant sacrificing her own well-being. "I've now realized I need to include myself in that sacred hoop, and not just talk about it. I'm in the process and exploration of what I'm calling 'healing justice.' We don't really use 'healing justice' very much up here north of the medicine line in so-called Canada. It's used a lot in Black communities, south of the medicine line in places like Atlanta."

She realized that the way of dealing with her own healing had to be different than what the dominant culture proposed. "Indigenous peoples really need to take notice of this and really understand what that looks like in our communities, in our ways of organizing. I'm starting to build this kind of understanding toward the pathway of healing justice, but realizing the ways in which I've been organizing and I was taught to organize by other older Indigenous activists and White activists that are just to 'do, do, do, do, do.'"

While she acknowledges the necessity of sustained action, Melina pointed to the need to take care of oneself amidst the work in order to stay with it long term. The alternative was what she had experienced in her collapse. "I was like, 'Okay, just die.' I was in so much pain. I was like, 'Okay, if this is it, then that's fine.' That's kind of sad that I had literally worked myself into the ground. There're no collective processes to support somebody that's just given to the collective for twenty years. So for me, that's what healing justice is, trying to understand how we bring back collective precolonial spaces, which have always been inherently Indigenous in a certain way. I don't want to co-opt the term 'healing justice,' but how do we bring healing justice into Indigenous communities in a way that we understand that collective healing and liberation is necessary, and needs to happen in collective ways and spaces?"

Melina came back around to ceremony, and its place in this discussion. In Canada, many of the ceremonies were prohibited until 1961, so everything went underground, something from which they are still recovering. She had also been studying Dr. Michael Yellow Bird, MSW, PhD, an Indigenous researcher working out of the University of Manitoba who is investigating how colonialism has affected DNA. Dr. Yellow Bird's work focuses on neurodecolonization and Indigenous mindfulness. Melina discussed how ceremonial practices and things like singing and dancing are also ways the body somatically releases stress and brings itself back into self-regulation. This is in contrast to the adrenal surges from activism being triggered and always feeling the need to respond immediately to things.

Like other Indigenous activists, Melina is regularly besieged with requests to run webinars, attend them, give speeches, along with countless other tasks. Her experience serves as an example for others in similar situations, after being bedridden showed her the impact to her body and spirit that years of organizing brought.

"Because of these [she held up her smartphone] and the computer, our bodies are at this high-frequency vibration that I don't think we can sustain. I need to find tools and spaces and places where I'm actually learning better ways of being and engaging with the world. Our people have already done this, through ceremony, fasting, through four-day vision quests, through lodge. There're so many different ways our people have had that regulation, of coming back into alignment with our body and with Mother Earth and the spirit world."

We asked Melina her thoughts on doing our work out of our obligation to serve the Earth and future generations.

"I think of our elders and how wise they were and are. They had reflection time. I'm not going to be a good elder if I don't make time for reflection. If all I'm doing is responding, I can't reflect, and I turn into someone I don't even recognize anymore. This is why the need for spaciousness is so important, and to know our limitations is a really important part of that."

Melina acknowledged the imperativeness of this, due to how reacting to things and always pushing forward, often frenetically, even into good work, generates the same energy as the frequency of that which is destroying the planet. If our work doesn't start with our feet being firmly on the planet, and our energy grounded in that way, it can, in a sense, make things worse. She has realized that if she does not do things in a different way, then she will not be able to tell future generations of the imperative to do things differently.

"I'm trying to do things differently and trying to model that because it wasn't modeled for me. We're trying to figure that out through the healing justice pathway that we're developing for Indigenous Climate Action and Sacred Earth Solar. I talk to our staff about it a lot because no one talked to me about this. It doesn't work to just keep charging forward while people tell you how strong you are and that you are a warrior, especially when sometimes you feel like you want to crumble and have a down day."

Melina believes the sacred feminine is coming back to the fore, and "rewiring the balance." She believes we are already seeing women-centered movements, and that younger generations no longer accepting the patriarchal-dominant way of being are signs of this rewiring.

We then asked Melina if she has been seeing a regeneration of rituals, ceremony, and things like potlatches in her communities, and what effects these are having among people. Aware of the immense complexity of healing generations of trauma, Melina has been acutely aware of the need to heal this after the Canadian government's attempt to annihilate their languages and ceremonies.

"We do see a resurgence, which I'm really excited about. We're in an Indigenous Renaissance. Yet growth is uncomfortable and messy. I don't feel like anyone has a blueprint for how the end result is. Each of us is always a part of the collective, and for me it's hard where I live right now since I'm not in my home territory. I'm in Vancouver, so I usually go to ceremony and lodge with the nations around here on the unceded Coast Salish territory. I think that's the beauty of cultural sharing, that I'm able to go into lodges here, yet things might be done differently."

Melina mentioned how their lodges in Alberta feel stricter to her, and she believed it was because their elders learned from their elders, who'd learned from theirs, who'd learned from theirs before that, so it was passed down in a different way. And regarding potlatches, she did not grow up with them, yet she appreciates the differences and linkages of Indigenous values that run across all nations.

"Like with the potlatch, we would call it a giveaway. Our giveaways are done during round dances. So we have a similar giveaway, but it's different. This reminds me of when I was in a potlatch at four in the morning, because their potlatches go so long; they started doing the giveaway and I was like, 'Hey, that's how we do our giveaway at our round dance.' If we have a ceremonial round, sometimes they're social and sometimes they're ceremonial. But there's always a giveaway, which is the redistribution of wealth. And I think that is what's lacking in an individualistic kind of Western society, which has a charity model that is not a redistribution of wealth. It's not about bringing equity."

Melina also mentioned the challenge of the global pandemic prohibiting her from going into lodge, so as a response she was finding her own ways of creating ceremonial space.

"I'm learning trauma-informed somatic understandings of breathing techniques, and how singing or humming our ceremonial songs can help my parasympathetic nervous system come back into realignment. I'm starting to understand the psychosomatic parts of ceremony for me, plus understanding how to listen to my system and body better. I became this disembodied person where I became very cognitive. I think a lot of us do that throughout our lives, becoming detached from our bodies. And our traditional dancing and singing and ceremonies bring us into alignment. I know of communities that have wailing ceremonies, where they actually let it all out, have this catharsis when there's grief around a death in the community, instead of just keeping it all in.

"Our ceremonies are the foundation of the healing work that we need to do. They are the foundations of how we connect to Mother Earth and the natural world. Then on top of that, because we're dealing with this system of colonization, we need to understand how to unpack and decolonize, in trauma-informed ways, which are sometimes ceremony or sometimes just old ways of understanding how we can bring our body back into alignment, with oneself and then with Mother Earth. It has to start there."

Melina's thoughts again highlighted the necessity of being grounded into Mother Earth before anyone can be of service. Otherwise, to come into service with the same frenetic energy as the dominant culture can be an exercise in self-defeat.

"Nothing makes sense if you're no longer connected. Maybe I will blame colonialism, because this system actively works against our connection, connecting to one another, connecting to self, connecting to Mother Earth. It does disembody. For a long time, I couldn't relate to the self-help/self-care frame because it was so individualistic. But understanding the impacts of colonialism and the real life experiences of trauma is necessary before healing can occur. It's not just about meditating. It's also about understanding, social liberation, and our collective liberation."

In closing, she maintained that doing her work must be in the proper context. The environmental campaigning, implementing of solar projects, shooting the *Power to the People* TV show about renewable energy and equal housing and food security, and the work to combat the climate crisis, are all efforts toward "bringing in the sacred view to a society that doesn't have that sacred view, which is why we have the climate crisis in the first place."

20

Medicine

COMPOSED BY STAN RUSHWORTH

Nothing is more important than the medicine. The one thing everybody agreed on, was that nothing comes before the medicine. The medicine is what holds us all together.

Our final interview comes from the high desert mountains of the Southwest, given by an elder who chooses to be nameless for reasons that unfold as he talks. He has a wry and open smile, an easy laugh, as he moves from deep seriousness to ironic humor easily, back and forth. He has been providing ceremony for people for many decades. Old hand-hewn beams carry the tin roof above him, and sometimes a breeze rattles the metal gently. It's a bright and sunny day.

My first questions are, "What is the medicine?" and "Why is it so important?"

He nods and begins. "I was taught young that nothing is more important than the medicine, and nothing, nothing comes before the medicine. The medicine was here before us. The medicine will be here after us.

"Some of us have been entrusted with a sacred trust to make sure that this medicine and this power is used for beneficial purposes for the people, to help them in their times of need and their times of stress and pain.

"To describe what the medicine is really borders on the impossible, because it's a concept. You're not born with it, you're born *into* it. And the elders around you help you to get in touch with everything, everything, you know?"

He pauses, reflects, then continues, the words rolling smoothly and defined.

"What is the medicine? What is the power? You can look at any of the major spiritual aspects that the people rely on or contemplate. People say there was the big bang, but what caused the big bang? People say 'In the beginning was the word.' What was the word? What is this power that causes life to move forward?

"To us, the medicine just is what *is*, and for it to really matter to people, in general, it's something that you have to be raised with, and if you're not raised with this concept, that we are connected to everything on this planet, in the universe, then as your life moves along, trying to wrap your head around that and comprehend the power that's available becomes very difficult, because people have been trained to be separate from this energy.

"For the medicine to work through the medicine people, to really truly work, what is essential is that the people who are involved, the people who need help, also understand what this power is about, what it's for, not necessarily meaning that they have the ability of tapping into it, but that they realize this energy exists. And so they put their faith, not so much in the individual, but in the energy and the power that this person sits in the middle of, between the power and the people.

"On that personally, what I'd like to say is that I am nobody. I have no name. I have no possession."

He stops to allow these words to rest in the air, to hold place, as bright sunlight moves across the earthen wall behind him.

"The reason for that is that in ceremony, the person who is running the ceremony is always spoken to in terms of their position, not their name. Intercessor, lodge leader. Every Indigenous people here on Turtle Island has this kind of title. And when we are in ceremony, the human aspect of who we are must be totally put aside.

"You cannot allow anything that you carry to interfere with the energy flow that's going on. If you do that, then whatever power is coming through to help will be short-circuited. That's a good way to phrase it, because you're still putting your opinions, chores, theories where they really don't belong.

"In that, I as a human being have absolutely no place to be. I can only be there in my position as an intercessor, and the reason you have to separate yourself is that you need clean flow of energy, clean flow of power.

"I know I'm not really answering your question very succinctly, because there's no clear way to describe what happens, to describe how it happens.

"There is a book by Vine Deloria Jr. entitled *The World We Used to Live In*, and he delves very deeply into how the medicine works, how people work with the medicine, and at the end, he talks to physicists who have discovered that 'Oh, you know, all you Indian people were really right.' There is a power that rides through everything. They call it by all kinds of different theories, but the fact remains that it is. It is. And that's it. It just is. And if you are raised from childhood to understand this, then this power becomes relatively easy because you know that it exists.

"When you've had generations of having that subdued through external forces that want to exercise power and control over you, then it's very difficult to understand what this really is, what this power really means.

"They keep asking, 'Well, what is it?' It's sort of like John Sebastian's old song, like trying to tell a stranger about rock and roll. You're just in it all the time.

"That fact that everything that we have around us, the trees, Earth, stones, each individual, is all part of this, is the way that it works in ceremony. For example, an elder intercessor who taught me described the sweat lodge like this: 'When you're sitting in there, it's like a flashlight. The intercessor is the bulb. The bulb can only burn as bright as the power of the batteries that surround it.' Every individual in there has to bring their power to this ceremony so that the ceremony is clean, the ceremony is strong.

"And if there are people in there to just take what's going on, then the power of the ceremony will diminish. Not drastically so that it doesn't work, but not a hundred percent of what's available. And a lot of that reverts back to growing up with the concept of what this power really does.

"What does it really mean? You know, all Indigenous people here have different ceremonies, but the end result is pretty much the same thing, which is to help the people who are in need, physically, psychologically, emotionally, and to help them the best that we can."

I think back to his comments about growing up in a relationship with this power, and I ask if he thinks it would be helpful to others to trace his life history as representative of his generation, from childhood with his traditions, through youth and other traditions coming in, and now as a mature man. I ask him how it's all developed over time.

"Let me start with this. In the beginning we *knew*, when a child is born. There was the introduction to the first sunrise. The elders, and usually back then it was the grandmothers, would take the child outdoors. And as the sun was coming up, they'd hold the child to the sun to get the power of life. The sun is life. And the words spoken, roughly, are: 'There is no power in the universe that is greater than you. There is no power in the universe that is lesser than you. You are one with all things.'

"When that kind of information starts at that age and just increases as you get older, your ability to comprehend what this power of the universe is becomes stronger, because it starts from the time that you don't even really have what they say is conscious memory, but you do have cellular memory. You have muscle memory, and everything around you is being bombarded by this power at that time. And then as your comprehension of language increases, the ability for the elders to pass on this information becomes greater.

"And in the old days—and I'm going to use Stanley here as an example— twenty-some odd years ago, he should have been able to stop doing what he has to do, the teaching and all that on an everyday basis in order to basically stay alive in this society. At that point, he would have been a grandfather, as he is now. He would have had children, and his children would be out there doing the work and the grandchildren would come to him, and he would talk to them about all of this. Whatever his path was, there would be certain children who would tune in to that particular aspect of their own being, and they would migrate to Stanley to learn what he knows and how he knows it. The other children, as they're roaming around to different elders, would find elders they were in tune with, and that's where they would go. Some of them would become great hunters, or fishermen, or trappers, or teachers telling a story. Some would walk the medicine path, but all of this could only be possible if the elders were there, free to speak to them without having to say, 'Well, okay. I can tell you this story, but I can only tell you half of it today because I have to go load a bale of hay.' The continuity is what's really important because in that flow, that's how we get into how the medicine works, and how it works as a flow is that you can't interrupt it.

"You know, things have changed. Back when I was growing up, if you were in ceremony—and again I'm going to use the term sweat lodge, because everyone theoretically understands it—if you've entered into the sacred place, and the door closes, from that point on you're in. End of story. There's no leaving, no breaking the circle. And if at any point in time somebody breaks the circle for whatever reasons, at that point in time, the ceremony stopped right there. It was done.

"There was no, 'I got to go out and do this,' and then coming back in and picking it up because once that circle is broken, that flow of the energy with the batteries, which are feeding the light, has been removed. So everything has diminished to the point where you may as well just close it out, because if you leave, you leave that sacred place, and you come back out into what they like to call 'the real world.'"

He stops at this point and looks out into the day, then carefully tells what can happen in this case, something to consider when choosing to come in. It's about the value of finding and holding the commitment, and consequences.

"When you come out of that ceremony, you're open and you haven't had a chance to fulfill the entire circle. You're open, so whatever's going on outside of that ceremony is going to enter into you, whether you're conscious of it or not.

"And then you're going to bring this back in, which is going to change everything that's been going on in there for whatever that time may be. One round, two hours, half an hour. All of that will change. Whatever it was you were going there to do has become irrelevant. So you may as well just put an end to it and shake hands, and as the old guys used to say, 'Let's go down to the house and eat' because it's done.

"Getting into that rhythm only really takes place and can only happen if this is the way you believe. I hate to use the word, but this is a belief. No, it's a *knowing*. It's a knowing that you are in a sacred place, a knowing that you're in a place of power, a place of medicine, and in that, it's knowing your ability to participate, because all of our ceremonies are participatory. Nobody just sits there and takes it in. They have to be present. They have to be spiritually, physically, emotionally present for all of this to happen."

He sits quietly for a moment, about to add something, but the phone rings nearby, and he shakes his head. "Please wait for that robocall to stop. I love the twenty-first century. So much for privacy, peace, and quiet." He smiles, then asks seriously, "Does that help at all?"

I ask him to continue talking about growing up during a time of scattering peoples, of war and tumult and social change. He nods, then begins with a laugh.

"Well, as Mr. JFK would say, 'I would like to say this about that.' You know, that's one of those things. It hurts my heart too much to deal with some days, to think of who we could have been, who we should have been. John Trudell says, 'We dream about a time that could have been different, but it wasn't, because we weren't.'

"But this is where it gets really interesting. There are people who, because of circumstances, will get a very specific series of teachings over their life because they still live in an environment where this is possible. Some of them hold on to it in a good way and make things happen. Some of them allow outside influences, which are not Indigenous influences, to bastardize what it is they do. And since, because of my people being scattered so much and having so many medicine people eliminated . . ."

He quickly moves to an aside here: "Historically, one of the first people they would always try to get rid of were the medicine people, because then they separate the elders from the children, separate your knowledge and your wisdom from the future. You make that split.

"So if you're not raised in that particular environment where you grow up with that, then you have to find a way to increase what's there. And so for myself, I spent a lot of years traveling around what people call the Americas, talking to as many of my own people's traditional people I could find who were still alive, who had bits and pieces of information, because the diaspora has scattered us to such a point that some of them would say, 'Oh, I know this song,' but it's a potentially long ceremony. They'd say, 'I know this song, and I know this part of the ceremony because my grandmother told me about it and showed it to me. But I know a guy, or a woman, who lives over there whose grandfather did it, so he probably has more information than I do.'

"So you can get a little piece of information, and you say, 'Thank you very much,' and you get this person's address, you get in your automobile, and you start driving. Somewhere between from where you're leaving and where you're going, little things turn up, like gas money, automobile repairs. This is for me, but I know other people who walk this path whose story is really similar in that way. You end up picking up a newspaper and finding work for a day. You end up picking apples in Oregon or driving a forklift in Wisconsin, whatever it takes. You work three, four days. You make X amount of money, you get back in your car, and you keep driving to where you have to go. This is where the information is, because the information is not readily available in the teepee, wikiup, longhouse, or on the other side of the village. You have to go find it.

"For me, one of the things that's happened over time is the people who've come later into my life. My time frame is the sixties, the seventies. Medicine, radicalism, you know, RedWind in California, and in South Dakota, the Yellow Thunder Camp. These places were being put together with medicine people who were there to teach. One of the first big ones was Red-Wind in San Luis Obispo, with Semu Huaute, a Chumash man. He put that together. After the Native American Freedom of Religion Act was passed in 1978, Semu was the first man to put a sweat lodge into prisons, which he did immediately. 'Okay, the bill's been signed.' He showed up, 'How you doing? Here's the law. I'm coming in. Build the lodges, and get the Indians to come in and sweat.' And before they knew it was happening, he had lodges in a lot of California prisons.

"RedWind is a real good example because there were medicine people from different walks of life. One of the people that I was really close to was Henry Tyler, who was Arapaho. At the same time, there were people there who were Lakota. There were Diné people, and some Indé [Apache] people, Phillip Cassadore in particular. He showed up there, and Mad Bear Anderson from the Iroquois Nation. He showed up there. Phillip Deere was there. I mean, there's a list of these people who showed up, and each one brought what their tradition was, knowing full well that so many of us have been scattered that we didn't have all of this information, and these men could teach. And there were women who came too. There's the Moves Camp family who came, who became really famous because of Ellen Moves Camp, and the Moves Camp family were very powerful medicine people.

"They came to places like RedWind to help us understand the power. 'How do you take what you have, in order to get into it? How do you make it work?' For example: I'm not a Sun Dancer, but I've been to Sun Dance, so what did I take from there? What did I give? What I gave was my support, everything I had, for people who were dancing. What I took with me was the understanding, through the power of ceremony, of the power they could bring in. The power of the pipe, for example, which I learned from Henry Tyler, Arapaho, more than I did from Lakota later. What is this instrument of power? What does it really do? How do you work with this in order to make things happen?

"In the old way, as I was told, people would wander and talk and meet other medicine people, and they would exchange information. So this isn't a new thing. One of the places you can really see this clearly is in New Mexico, with the Jicarilla, or with Diné and Apache people. The influence is very plain to see, as in the Kiowa-Apache, where there is a combination of the two. So gathering the medicine from various walks and putting it together is not a new thing.

"No, it's really old, and one of the problems we run into is, 'This is the only way!' Well, I don't know about that. I sing this song, and while it's not traditionally my people, at the same time, there is the power that I've picked up from a singer who sang this particular song, sitting with this person and singing it with them, over and over and over again, until they were satisfied that I actually understood what this song was supposed to do, and said, 'Okay. You got it. You can go sing it now.'"

He emphasizes that it's not about the number of songs you know, but the relationship to them. "How many of these songs know you? It's a symbiotic relationship. The spirit people have to work through someone to make things manifest. And since you're on this side of the line, you need the power of the spirit people to actually do the job that there is to do. You can't separate the two. That's just impossible."

The word "separate" brings him further into how relationships work, in the song and in daily perceptions, so he steps into this connection for a moment.

"I hear people talking about 'the environment,' like it's a separate thing, and it's not. We are the environment. We are the sum total of everything that's on Mother Earth and there's no way around it. We drink the water, we eat the food, we breathe the air. Whatever we are is the environment. If you treat 'the environment' negatively, obviously you treat yourself negatively, but you just don't realize it because you have been taught that it's a separate thing. So it really is about the ability, from what you're taught, to be in the wilderness, which to us was never wild. That was just home.

"You know, there's a tree like the oak, a strong tree. You cut it down and the fire is always hot and it burns long. On the practical side, this is true, but on the spiritual side, what does Oak represent? It represents stability. It represents deep roots into the Earth, and longevity. There are all the spiritual aspects of this being that should be understood that aren't, and it's not understood because people have been separated from that part."

Coming back to the camps and their teachers, he describes how they worked for those who had been deeply affected by the removals of people and traditions, and what they offered toward healing that separation. This is the same separation everyone is suffering from today, and I recognize names of people who gave warnings half a century ago, like Phillip Deere. The camps and teachers were working for everyone's long-term benefit.

"At places like RedWind, you would find people from different nations, and because of who was there teaching, people would start following not necessarily their own traditional road, because their road had been totally severed, totally scattered. A lot of them ended up following Northern Plains traditions because these people held it together for a long time, and they still do. You found people coming down from there, like Selo Black Crow. I know people who are Huron who are Sun Dancers. I have a good friend who's Lenape, and he's been doing this twenty-five years. What I'm trying to say is that it's not 'It has to be this.' If this is the path that you're on, the path that you follow, stick to it and move forward with it. And take what you can from each one, because they all know.

"And again, things have changed. Once upon a time, the Sun Dance was done by one person, but because of the way things move and change, it became two or three people. Now you have a bunch of people at the tree, and it doesn't diminish in any way what the ceremony is. It's just grown because the need to have this happen has grown. It just has to be this way. Because we live so far away from our natural way of life, we have to regroup within ourselves, recharge ourselves in order to do something simple, like walk out the front door of the house and go to Walmart and deal with a whole bunch of people that have no idea who you are, where you come from, and don't want to know. And personally, I'm okay with that. I really am."

He smiles and looks out the window toward the sun, and waits, thinking. I wait too, then ask him to continue with the reasons the medicine has been endangered, how and why it was attacked, lost, and broken, for this is our backdrop. His expression does not change as he continues, beginning with the boarding schools.

"The people who instituted this whole process were smart. The equation is all messed up, but they're really smart, and they've been doing this since Constantine." He shakes his head.

"It's important to realize where this mind-set started for these people, and how through coercion more than anything else, it spread to the point where the people who were running the boarding schools really believed that what they were doing was right. 'This is the right thing to do because, you know, we're good Christian people.' And of course, if you know anything about Christ, these people are anything but Christian, but that's up to the Jesuits to fight over." He laughs, then becomes deadly serious.

"They said, 'We're gonna kill the Indian to save the man.' But that's impossible. So a lot of children who were sent to boarding school just died. They had heartsickness, homesickness. They were separated from who they really are, and they just quit living." He stares as he finishes the sentence, then waits for more words.

"The people who instituted this in the beginning could see exactly what they were doing. Again, if you separate the children from the elders, you're separating your knowledge from your future. And the separation gets bigger and bigger. If you take a kid five, six years old, and you take him for the next eight years, you subjugate who he is inside to become somebody else. And the majority of the time this really worked. They got so beaten down that they started to believe what was being poured all over them.

"But then there were the other ones, who were worked over by the school and said, 'Is that all you got?' 'Well, if you don't do this, we're going to beat you black and blue.' 'Yeah? Go for it. I know who I am. I know where I come from. You can't take that away from me, no matter what you do.' There were those who went that path, and I'm happy to say that's my family. 'This is who we are and that's just not going to change. I can't give it up.'

"When you look back through your people as an Indian, any of us, back through our history, with the sacrifices that were made just to exist, how can I do any less? In the time frame and world that I live in, how can I do any less than what these people did?

"Leonard Peltier says in *Prison Writings: My Life Is My Sun Dance*, 'I look back at what my people went through 150 years ago, a hundred years ago, 50 years ago, at the sacrifices that were made, people standing there watching their children die in their arms, just so I can be here, to sacrifice some more. I will sacrifice, and I'm not leaving this place until my people are free.'

"And what happens is that people don't really see that. He is in prison for forty-five years, and you have the law and the reasoning behind it, but *why* did this happen? Leonard's in prison for his entire life because he is an Indian who wouldn't kiss Uncle Sam's ass. That's really what it's about. They didn't know who did it, but 'Leonard's here, so someone's going down. I guess it's him.' Equal justice."

With this rule of law, or "law of rule," he is impatient, and he follows a rattling gust of wind on the roof with a plain fact more people should know. With this fact, he returns to how his life reflects that of many men and women his age, our purpose today.

"We had our freedom of religion declared illegal in roughly 1890. We were not given it back until 1978, so we had to hide much because of what was going on. And a lot got lost because people didn't put out the energy to seek it, to become part of what this circle is.

"And then they opened the door and brought in non-Indians. And Indians became this too. I don't like to use the word corrupt, but corruption has happened, and people don't take the time to really learn what's going on, from the beginning, you know, like how to stand at the fire pit."

It's about pragmatism for him now, as he describes his own path alongside that of others. "No matter where else my life has gone, I've always considered myself a fire man, because it's the beginning of everything. With no fire, there's nothing else that's going to happen. You go to the Sun Dance, they light the fire. You go to the sweat lodge, they light the fire. You go to the kiva, you light the fire. Always the fire. The fire is life. It's one of the first big gifts we got, you know? This is so important, realizing that the fire itself is a living, breathing being. It's alive, and you as a human being must participate in its creation."

His eyes are bright as he describes this, showing a deep satisfaction and joy that covers years of experience with the practical nature of making this deep connection.

"If you want to have the ceremonies, 'How do you make this fire?' It's wondrous. And in that building of the fire, 'What kind of wood do you use? How do you build a fire pit?' This is a sacred ceremony unto itself. Not all the fire pits are the same. It depends on whatever tradition you follow. How they face is different because of how you view different things. And none of them are wrong. When people say, 'Oh well, it's like this . . . ,' it brings up the old joke, 'How many Indians does it take to change a light bulb? It takes a hundred, one to change the light bulb and ninety-nine to say, "We don't do it that way."'

"And therein lies a huge problem, with saying 'We don't do it like that.' Well, it doesn't make it wrong. It just makes it different. We have to learn to take the time to learn, and once you learn one way, whatever it may be, just keep working that until you become one with that ceremony and the ceremony becomes one with you. Again, it becomes a symbiotic relationship. You're the same thing. You're not just digging hole in the ground, filling it with wood and fire and rocks. You will come up with hot stones, but they won't have anything to tell you when they come into the lodge, because they haven't been fed in the way they have to be fed."

In this process of learning, and dealing with the scattering of traditions, and with living in another way of life, like the Walmart he mentioned earlier, I wanted him to continue with how his life touches his peers' lives, and with what matters in this light.

"In the book *Lame Deer, Seeker of Vision*, he talks about medicine people as human. The only way he can help anyone looking for help, is to have experienced what life is about. I cannot sit there and contemplate pain, misery, agony, separation, from just a concept. It's something you really have to experience.

"So the more you learn about yourself, when people come and ask for advice, you lay out options that these people haven't seen yet, basically because they're thirty, and you are seventy. And in those forty years, you accumulated a whole pile of options that you can lay out for people to look at, and you're in a position to discuss what these options are, what they really mean and how they can apply them to whatever the situation is.

"Lame Deer said people don't go to medicine people because they're good, but for power, so once they enter into this arena and come into earning *not* their title, but what they're supposed to be, they have to do good work. And that doesn't mean that they're good guys all the time, because they live in the real world.

"But when they enter into the ceremonial state, all the other stuff that they may or may not be 'guilty' of ceases to exist because they step into a different realm, and everything they've been doing in the 'everyday' realm is what is actually going to help them deal with other people going through these problems. They can say, 'Yeah, I did this.' Because they've been through it, they can open it up. 'Okay, you know you're going off track. Why?' You know that walk, so you walk them through it and say, 'There are ceremonies we can do to help you with this.'

"The Diné have tons of ceremonies. They have singers who do very specific things for specific reasons. And we understand that medicine people, even the ones who've done it their whole life, don't have all of it, but they are very specific in what they do because it takes a long time to become proficient, to be able to get into that zone to make things happen.

"There are people who say, 'I have two ceremonies.' And in those two ceremonies, you're looking at thirty or forty years' worth of life just to get those two down. The ceremony is so intricate that it takes a long time to really master, going into that place and allowing that energy and being able to separate."

He comes back to the word "separation" again, but this time it's about differentiation, an essential part of the process.

"What you think and what you hear are two different things. The spirit people talk to you, but if you're still carrying some of that 'managing' with you, there's a confusion between 'what I think,' and 'what the spirit people are saying to me.' That cannot overlap. Whatever it is you're thinking, you have to leave that 'out here.' You have to come in clean with no extra, and with no expectation of what's going to happen. 'This is what can happen, but will it happen exactly like this?' Probably not. In the universe, everything is in flux. Everything is in movement, and from the time we start to the time we finish, things are going to change."

As he continues, he quickly and steadily moves through the many things to consider in what the world has become, and what needs sorting, differentiating, alongside what needs to occur.

"Things will change, and if you're sitting in the intercessor's position, as we used to say back in the sixties, 'You just have to move with it.' And, as we've talked about, it's also created by everybody who's in this ceremony. They have to bring something to the table; even if it's the simple belief that this is good, they have to bring something to the table."

It's not easy in today's world, and he goes into this. "In the world that I live in, we only get X amount of time that we can really focus on what's going on there because there are little things like 'I got to pay the rent, I got to pay insurance.' I have to go out into the world and communicate with people, and normally I'd be okay with this, but I've been given an offering to do a ceremony. So I should be able to spend all of my time focusing on the ceremony. Period. But I can't do that because all this other hustle is going on.

"I cannot avoid it by packing my car and going up into the mountains and sitting there for four or five days while I get ready to do the ceremony, and then drive back down and get ready for the ceremony. Because in those four or five days, little things might happen, like maybe a pipe broke. I ran out of propane. All of these everyday things have to be taken care of, so that actually carves into the time we have to do ceremony. That's why, like I said, in the old days, when you were growing up you took care of all of these things. You became responsible for the well-being and the keeping of your elders, your children, the women, the grandparents."

Contrasting what we deal with today, he described it as a "whole system that was in place," and that the teachings "came to you," rather than you chasing them all across the country, filtering it all through a world from the outside that is influencing you to "civilize." But even with those influences, a deeper impulse still exists.

"Even there it's interesting, like with education. My grandfather went through the boarding school system, and while there, he met Indians from all over the country. There was no way around it. You picked them up from wherever they were from, in Arizona, or Florida. You stuck them someplace, and part of that is to separate them from the people.

"But the other side of that is that finally, as we discovered later on with the relocation program, they find five or six different nations ending up in the same neighborhood. And so you have Pan-Indianism, which isn't really a bad thing because at least if nothing else, you have a strong cultural base.

"When these kids went to boarding school, just getting through it was the best thing you could do. But they also built friendships with people who were 'the traditional enemy,' whatever that's supposed to mean. All of a sudden you find a bunch of Diné getting along really well with a bunch of Hopi. You find a bunch of Apaches actually getting along with the Zuni, who gave them the name 'the enemy.'

"Because we're all stuck in this place, and the only thing we have in common is that we're Indians, we may as well join together. I saw this as I was growing older because of where my grandfather settled, when all these other Indians came because they'd gone to boarding school together.

"And on the road to this, after they'd left World War I and World War II, they went back to Los Angeles and back to their reservations, and already that early on, in the 1920s and later, things had changed. They could see the corruption coming in, the alcoholism, the abuse. You could see it coming out of the boarding schools.

"So if you're going back to the homeland where things are supposed to be peaches and cream, you see they're not. And they hear where the kids are that they grew up with, so they pack up and move to where these people are, so they can be with friends and not have to fight the bureaucracy every day. And also, to not deal with the 'handout,' because a lot of the people would say, 'Well, they give us this.' 'Really? Well, I can't live like that. They don't own me.' Some would say, 'It's great that I get a check once a month and some food on the table,' and others would say, 'That doesn't even cover the interest, never mind the principal.' You get a check for eight hundred dollars a month, and it costs you half of Oregon and a piece of Idaho, so it's okay? No, it's not okay at all. These checks just make it look good as you're going down in flames.

"And many of the men who'd been in the schools saw this when they came back from the wars. There was a group out of the Midwest back then called the Red Arrow Brigade. They were actually among the first troops that went to Europe in World War I because all the schools then were so militaristic in their training that this group of young Indian men were better trained to be soldiers than people in the army. So they shipped them over there real early, to deal with that. And in that they got to see more of what the world is like out there. Good, bad, and indifferent. They got to see it all.

"And so when they came back here, they had a broader understanding of what's going on out there, and to see that this is really not right, the reservation. My grandfather always called it a prisoner-of-war camp. And that's all there was to it, to him. May he rest in peace.

"There were a lot of these guys who felt that way, so they moved back to where there was farmland. And with that, when I would be with my grandfather and he'd be teaching me, I was also learning from people who were Dakota, Lakota, Nez Perce, Hopi. They were all there because they went to boarding school together, so the influx of information came from many different people. But the thing that tied it all together was the fact that the one thing everybody agreed on was that nothing comes before the medicine. The medicine is what holds us all together."

He stops now, then slowly begins again by detailing the nature of relationships today, and how this connects with his past, and with those men and women his age, continually moving through important contrasts, reflecting on changes he sees.

"We are a people of spirit, but a lot of people get hung up with, 'If you don't speak the language, you can't do the ceremony.' Actually, as far as communication is concerned, the aspect that is essential is that you understand the songs, because they have tonalities and energy, and through this you totally understand what the songs are, what they really mean.

"But if you're going to be speaking to a larger ceremony situation with groups of people, then even if they're all Kiowa, for example, they weren't all raised to speak the language. They weren't all sent off to that particular reservation. These people flew in from New York, Chicago, Minneapolis, and they show up because they want that connection to their home. They want their connection to the people."

He describes attending a gathering with an elder of his, in which a man talks for a very long time in his language without any translation. "I need to know what it means because I'm standing and honoring what's going on, and I'd like to know what I'm honoring. 'Okay. If I don't know what you're talking about, I'm going to sit down.' Well, this guy goes on and on, and my elder, God bless him, leans over and whispers in my ear, 'And now he's blessing his shoe laces.' And it's only half a joke. An old woman was standing there and she just sat down.

"She was a local California Indian. I can't remember which nation, but she was wearing the woven basket hat they make, and with the clapper stick. I think it's all cool to see, and that I can walk up to this person and say, 'Hey, how're you doing? What does all of this mean? You build a hat like this, a dress like that,' while we do it this way because of this reason. And so it becomes an education, and guess what? We're communicating in English because that's the common language we have, for better or worse. I always find it interesting that we were given this common language, and we don't use it to communicate with each other. We use it to put each other down."

He shakes his head at that, then waits quietly. This brings us back to the essential separation again, the division of people from people, from Earth and everything on her, so I move toward the conclusion, asking "What's really at stake here? And, what do you have to say to young people regarding the medicine today and into the future?"

He nods, then gets up and moves outside, with some physical difficulty, chuckling and saying, "Everything used to be easier." The sun is bright, and as he settles, I comment that these final questions are closely related.

"They definitely are," he smiles, but is very serious.

"Well, number one, they don't have to keep the medicine alive because whether we're here or not it will exist, but it's about the utilization of it to make life better, and that requires a commitment to a way of life. I don't know if I should say it's the antithesis of the world that we live in, but it pretty much is, in a lot of ways. There's no profit margin, right up front. You can put in all the years of study, the years of practicing, you know, because they can only take you so far, and then you have to just go out and do it. And there's a sacrifice, to get that which is beneficial for human beings. So for people who need this connection, even if you're not walking the medicine path, you're relying on that to get you through life. And you have to *be* what you commit to in the simplest things. We wake up in the morning. Do you have your own little altar? You say a morning prayer, 'Thank you for another day of life. Thank you for keeping me pure during the night. Help me to be a good human being.'

"It's a practice. And as you go out today, little things will happen. Maybe out of nowhere your foot slams on the brake, and some car is blowing through the intersection and you knew it was happening. You reacted. Something said, 'Do this,' and you did it. Do you shout out the window or give thanks to the spirit people who put that little bug in your ear? 'Thank you for watching out for me.' That's all being created by the medicine, the power that flows through all things.

"My grandmother said, 'You opened your eyes this morning. It's the first miracle of the day, just like the parting of the Red Sea or walking on the lake.' Those are the big ones, and for those who've been in the lodge, every now and then they have manifestations, and people say, 'You must see this all the time.' If I saw this all the time, there would be no reason to be here because I'd be back in the spirit world. This manifestation is special."

He stops and contemplates the question of what's at stake again, emphasizing the key idea of what makes everything work, no matter who it is or what's going on. He keeps young people in mind as he builds his thoughts.

"This is really important, that in order to keep the 'medicine' alive, it is something all of us have to practice, not the same way that ceremonial leaders would, because that's a whole other thing, but it's just part of your life.

Again, you get up in the morning, say your morning prayer, you give thanks for the day. I do my morning prayer before breakfast, and I do my evening prayer before dinner, and when I'm doing my evening prayer, I'm asking for help and guidance through the night that's coming.

"That's my thing, and everybody does something different. I'm not here to tell you this is how you have to do it. There is no set prayer. It's not for the show. You're talking to the Creator. I just think it's important that people get into that rhythm, that communication with the spirit world, with the power, with the medicine, and attend to their lives as though they were given many gifts. For some of us who've been through all kinds of crazy things, that we're still here is obviously a gift from the Creator, to do whatever it is we have to do, but always making sure we're doing it in a good way. The best way we possibly can.

"For the young, some words are out there, but I'm going to have to go back a little bit. I spent many years on the road, learning, doing ceremonies and learning more versions, and I was always on the move, which made it difficult. But it's a good thing too because it gave me a chance to learn, and because I was in motion all the time, it also gave me an opportunity to pass it on. You're only there for a short time. You end up sleeping on other people's floors, sleeping on the couch, and you do this and move along."

He stops, thinks back, traversing time, connecting an early lesson with the present, and with a monumental shift toward another balance in his life. "Somewhere along the line as a child, I also visited my aunt, so I was taught women's medicine, which really freaked me out at that point in time. But my aunt said to me, 'Look around. Who else is here?' and then she said, 'Don't worry. At some point in time, a woman will come into your world and you'll never have to deal with this.'"

A big smile crosses his face, with a light in his eyes at the change predicted by his aunt. His expression is one of deep gratitude, more than he can say. "Well, I was lucky. It happened," he says. The words carry a lot, and he moves quickly to his point. "It happened, and so now when the young women or the older women show up for women's situations, I can say, 'Hello. I'm going out for a cigarette.' Because she knows more about that than I do.

"What also came along with that, and her sacrifice for this is much more than the sacrifice of being on the road, was a stability. It could all be in one place. We could pass things on, understanding how this works and why it works that way. And in order for this to happen, this woman actually went out in the world and took care of everything, so that I would be free to do this work. And she did it not for me personally, but for the medicine.

"If you are in a situation like that, you ask, 'What did your partner sacrifice in order for you to make this road really, really work?' And you both did it, so it also works two ways, and when she was ready to do ceremony and make things happen, then it was my place to make sure that it all went well for her, to take care of whatever had to be done, and then step out of it, because that's not my realm. And with this stability, the door is open for the young to come and to learn and to study how this works."

He waits a moment, then reflects on his experience with that open door, observing the effect of the surrounding society on the time it takes to learn, to really dive in deeply. And he observes other priorities too, something to look at closely.

"At the same time, I've talked to other people who walk this path, and we've come up with this: Yes, granted, they get involved on the spiritual level. They learn, but there's always something else. 'Well, I'm gonna get my degree in anthropology.' And they put a lot of time, energy, and effort into that. The spirit path becomes the part-time thing."

He pauses, looking down, then looks up and goes on. "It's not a full-time focus, and I think a lot of that comes down to 'the stuff.' If this is the path you're walking, there's not a whole lot of 'stuff.' There just isn't, you know? And you come to grips with that very young, or not.

"I always consider myself fortunate growing up where and when I did. We didn't have anything, so I just needed 'enough.' And when you walk this path, it's a total commitment. There's no 'something else to do.' You may do something else on the side in order to maintain what you're doing here, but the main focus is the medicine path."

He stops, and his face softens, in respect, because he is not discounting what people feel they need to do. It's simply an observation of life, but he is clearly qualifying what he feels is needed for *this* path, and the consequences of what he sees. It's not a simple thing.

"Getting an education in the White man's world out here has its benefits, but the other side of that is just spending more time with that and not really jumping into the spirit path. I mean, I don't have as much of these ways as my grandfather, and he didn't have as much as his father, but these people will have even less because they have not been putting time into doing this.

"If there's a ceremony that's going on, then what you want to do is to find out who's doing the ceremony, and where it is, then go there and watch, even if it's something that is different from what you carry within your ways, but you go. You watch, you feel it. You don't go in with any preconception about what's happening. You go there and you observe and you feel what's going on.

"You have to go. You may only stay five minutes and then leave. I got that, but you have to show. When you're starting off, you have to show up, listen to what the people conducting the ceremony have to say, and also listen and feel the people who are there. Are they walking this path or are they tourists?

"If the young will step into that arena to find out what's happening and how it's done, then the next generation will actually get even more. Then my generation, your generation, will get more because it will be from a stable point of view, and it'll start young. It will start.

"A lot of it comes down to parents saying, 'I'd like to bring my kid, but he's going to go play softball.' Yes, I did all of that too, but at the same time, I'd tell my friends, 'My grandfather's taking me up to Maine to meet people and we're going to do ceremonies.' I'd go to ceremony, and they'd go play baseball, which is why I'm a one-blanket Indian and they have a Mercedes-Benz. And I'm happy with my blanket because it's my blanket, my blanket. And it was a gift that came to me."

He laughs at the Blanket vs. the Benz, and clarifies that it's never about money for him or for any of his teachers or family. "Somebody along the line says, 'Yes, I'll show up to do ceremony. I'll do that, and it's going to cost you a thousand dollars.' So did they come down with a bill from the Creator that said, 'You owe me $850, and you can make $150 profit?'

"All that doesn't happen. This was given, a gift. And any gift that you give back, you explain to people, 'If you want a ceremony, when you show up initially, there's an offering, and you do something organic. It could be tobacco, sweet grass, pollen.' There's all kinds of things that get passed on, and eventually they will end up in a ceremony. And when it's over, you have the option of gifts, and in the gifting, maybe the gifting is enough to put gas in your car so you can get from Omaha to Tahoe, and when that usually takes place, it's very quietly done. You shake someone's hand, slip them some money as they get ready to change their clothes at the end of the ceremony. Or they pick up their backpack, and there's some money stuck in it. Or there's a gas card stuck there underneath their towel. And that's really all that's truly necessary."

He stops, smiling and shaking his head. He wants people to know how important this is, that it's not a matter of exchange but of respect, and nothing more. It's a caring, and his face reflects this.

Thinking of where it can all go from here, I tell him about a young man fourteen years old who picks up an eagle feather by the lake his people have lived on forever, and he hears a song. I tell him of another man who as a youth is taken to the edge of the sea by his elder and told to strike rocks together until he hears the ancient song. Still another tells a youngster to look into a piece of driftwood until he sees a mask emerge, then to carve what he sees.

"Yes. The AIM [American Indian Movement] song was the dream of a young kid. He dreamed it. The thing about that is that you have to allow yourself to be open for this to happen. The young man goes to the lake where everybody's gone for centuries, but somewhere within himself he's willing to sit in silence, whether he got it from a grandparent or the way his people live, I don't know, but he sits in that silence so he can hear, hear what's going on."

He chuckles and adds what Graham Greene says in the film *Thunderheart*: 'Listen to the wind, Kola.' You know, you have to listen to the wind. It speaks to us. And you have a grandfather who says, 'Look at this wood until you see what's in there. And if you shut off everything else, the picture will appear because it's already there.'

"When you're spending all your time trying to learn, all you have to do is remember, remember the people. You have to remember this because it's inside all of us that way. It's there, and it gets passed on, what's necessary.

"What was it that Einstein said? 'I don't teach. I just provide an environment for learning.' And that's what we do with this. You can show the ritual, how things are constructed, but the actual application can only come from doing it. These particular young people are in an environment and a place where they can do that, and places like that should be made available for young people to do that, to sit, to go into the quiet. Go to the ocean, look at the wood.

"Out here, I love to sit up in the back of the property and just look up at the mountains, and they talk to me. Information has been coming through to me about old things that happened, and this comes through just from being on the land where they happened, sitting on it in quiet and letting it vibrate through. People have got to take the time to do that. This is all going back to plugging into what *is*, because it's there forever, you know? And we can plug into it if we separate from the noise, get out of the noise.

"The deeper the civilization, the louder the noise. There's just too much going on. Even the simplest thing, like electric wires going over, is enough to upset the rhythm. So you find some place where that quiet exists and you sit and sit in it. And it's funny because in those places, you're not really thinking. You're feeling it. You're letting it come through, and you can't argue with what you feel, you know?

"When I was with my grandfather, he said, 'Your mind is like a toolbox. When you need the tool, you open the box, get the tool out, do what you have to do. Then you put the tool away and close the box and live with your heart.'

"You feel with that energy flowing through you. And I think it's wonderful that these young folks are doing that. We need more of them, moving on, so that the next generation after these kids will have a better opportunity and an understanding of what this really is.

"When I was growing up, my uncle used to take me out into the 'wilderness' and have me sit there all day, and we wouldn't talk, just sit there, and it took a while, but once I got into it, I could hear the rustling of the breeze going through the grass and the leaves and watch how the leaves move, and watch all of these things going on. And then he'd put a blindfold on me, so I had to use the rest of my senses. Smell. Feel. Taste. You go through all these processes, and your connection to all of that becomes stronger because you have actually been in it, participating in what's going on. And he would do that to me at night, and I was just better at night than I am in the day. I got that moon energy, and am stronger there than in the day.

"And most of the old ceremonies we actually do, like sunrise ceremony, were done at night. Night's always a good time for that because it's quiet. It's very quiet, and the only thing that's moving at night are the night creatures, and they speak to you."

He smiles, laughs, then brings in another movie classic for his generation. "It's like in *Powwow Highway*, when Philbert's sitting in the lodge and he goes into that zone and you can see the manifestation of White Cloud walking up to him and touching him with that arrow, and Coyote is there. Is it real or illusion? You have to figure that out. You have to say, 'This was real.' It wasn't a figment of your imagination. This really happened, being touched by the spirit people, in order to move forward in the world that we're living in. I think that people are touched by this more often than they realize. They just don't pay any attention.

"And that's what this is about. It's here to speak to us. But who's listening? When I was younger, I used to go up into the mountains, and find a place, you know? 'This is the place.' I'd set up my camp and sit there. Later on, people would show up with tents, carrying enough equipment that they may as well stay home with the TV. They'd come in and set up camp, then they'd get up in the morning, have their breakfast, pack up, and leave. I'd ask, 'Excuse me, what's your rush?' 'Well, we have to go see this. We have to go there and see that.' 'But it's right here in front of you.' 'Well, I saw that, and now I'm going to go see something else.'

"To actually see it is the catch. Yep, that's a mountain. But you know that mountain there and the one that's just off to the side? They have nothing in common but that they're tall. I go to Silver City, up over the pass, roughly a two-and-a-half-hour drive. It takes me three, sometimes four hours to get over that pass because I go over at different times of the day and the sun is different. The shadows are different. Everything is different. The trees are all different, and 'Oh, wow. I never saw that before,' so I pull over and stop at the side of the road and look at it and watch it. Every time I go over, I see something different.

"So sit for a day or two or three and just chill out. Actually become part of the nature that you're in, because that's who we really are. That's who we really, really are. All the rest of this stuff is an illusion. It comes and goes.'

"History can tell you civilizations come and fall, but the dearest gift of all is here, and as John Trudell once said, 'We cannot destroy the Earth. We can only destroy our ability to live on it.' And that's what's happened here. People destroy it, and are very close to destroying our ability to live on the Mother. And that's sad."

He looks down at his hands, staring at them for a long moment, and then he looks back up sharply but quietly, secure in the words he finds next.

"The kids, the young. They need to do that. They need to get in touch with things that way. And if they're exposed, or allow themselves to be more exposed to all of it on that level, there will be a few, because it's never a lot, who will to the best of their ability walk away from this civilization and walk that medicine path, because they'll realize it is forever.

"It's not transitory. It's forever. And it starts with doing that prayer, that real communication in the morning and the evening. Make that connection and listen, listen, listen to what the spirit people have. 'Listen to the wind, Kola.' Wind will talk to you. You know, it carries everything that's going on in the world, so pay attention."

He smiles, shrugs his shoulders as the breeze and the tin roof talk together again, then concludes with a chuckle, "And what else do we really *have* to do anyway?"

Acknowledgments

We thank all those who have held strong over the generations, resolute in keeping the ways of being alive that are in these pages.

We thank all the participants interviewed in this gathering, for sharing these perspectives, feelings, and courage. You are this book.

We thank those who reach out to read these thoughts and experiences, for voices need listening ears. You too are this book.

We thank and acknowledge all the families, friends, and extended families of those who have given their support for this writing in every way. This includes: the Lannan Foundation in Santa Fe, New Mexico; The New Press; the representatives of John Trudell, who allowed us to use his line for our title, *We Are the Middle of Forever*; Linda Yamane, whose invaluable work graces the cover; and the Center for Imagination in the Borderlands Matakyev Fellowship at Arizona State University, for their generosity and encouragement.

We are grateful for this matakyev.

About the Editors

Dahr Jamail is the author of *The End of Ice: Bearing Witness and Finding Meaning in the Path of Climate Disruption* (The New Press) as well as *Beyond the Green Zone: Dispatches from an Unembedded Journalist in Occupied Iraq*. He has won the Martha Gellhorn Prize for Journalism and the Izzy Award. He lives in Washington State.

Stan Rushworth is a teacher of Native American literature and the author of *Sam Woods, Going to Water: The Journal of Beginning Rain*, and *Diaspora's Children*. He lives in Northern California.

Publishing in the Public Interest

Thank you for reading this book published by The New Press. The New Press is a nonprofit, public interest publisher. New Press books and authors play a crucial role in sparking conversations about the key political and social issues of our day.

We hope you enjoyed this book and that you will stay in touch with The New Press. Here are a few ways to stay up to date with our books, events, and the issues we cover:

- Sign up at www.thenewpress.com/subscribe to receive updates on New Press authors and issues and to be notified about local events
- www.facebook.com/newpressbooks
- www.twitter.com/thenewpress
- www.instagram.com/thenewpress

Please consider buying New Press books for yourself; for friends and family; or to donate to schools, libraries, community centers, prison libraries, and other organizations involved with the issues our authors write about.

The New Press is a 501(c)(3) nonprofit organization. You can also support our work with a tax-deductible gift by visiting www.thenewpress .com/donate.